ARCHAEOLOGY
UNIVERSAL LIBRARY

ANTIQUITIES OF ATHENS
AND OTHER
MONUMENTS OF GREECE

雅典古迹

[英] 詹姆斯・斯图尔特
[英] 尼古拉斯・雷维特 著

华东师范大学出版社
・上海・

图书在版编目(CIP)数据

雅典古迹 = ANTIQUITIES OF ATHENS AND OTHER MONUMENTS OF GREECE : 英文 / (英) 詹姆斯・斯图尔特, (英) 尼古拉斯・雷维特著. — 上海 : 华东师范大学出版社, 2021

(寰宇文献)

ISBN 978-7-5760-1493-8

Ⅰ. ①雅… Ⅱ. ①詹… ②尼… Ⅲ. ①古建筑 - 建筑艺术 - 希腊 - 英文 Ⅳ. ①TU-881.545

中国版本图书馆CIP数据核字(2021)第051769号

雅典古迹

ANTIQUITIES OF ATHENS AND OTHER MONUMENTS OF GREECE

著　　者　[英] 詹姆斯・斯图尔特　[英] 尼古拉斯・雷维特
特约策划　黄曙辉
责任编辑　王海玲
特约编辑　许　倩
装帧设计　刘怡霖

出版发行　华东师范大学出版社
社　　址　上海市中山北路3663号　邮编 200062
网　　址　www.ecnupress.com.cn
电　　话　021-60821666　行政传真　021-62572105
客服电话　021-62865537
门市（邮购）电话　021-62869887
地　　址　上海市中山北路3663号华东师范大学校内先锋路口
网　　店　http://hdsdcbs.tmall.com/

印 刷 者　广东虎彩云印刷有限公司
开　　本　787 × 1092　16开
印　　张　18.25
版　　次　2021年4月第1版
印　　次　2021年4月第1次
书　　号　ISBN 978-7-5760-1493-8
定　　价　380.00元（精装全一册）

出 版 人　王　焰

出版说明

詹姆斯·斯图尔特（James Stuart）是苏格兰建筑师、考古学家和画家，绰号“雅典人”，是英国艺术史上“新古典主义运动”的关键人物。

斯图尔特出生于1713年，年轻时就展露出绘画天赋。1752年，他前往意大利担任画师，学习拉丁语、意大利语和希腊语，并研究意大利的艺术和建筑。在那里，他遇到了尼古拉斯·雷维特（Nicholas Revett）。两人结伴前往雅典考察建筑遗迹并出版了《雅典古迹》。

《雅典古迹》共4卷，第一卷出版于1762年，其后分别于1789年和1795年出版第二卷和第三卷，最后一卷出版于1816年。1841年出版第二版，1858年出版第三版。《雅典古迹》还被翻译成法文、德文和意大利文。直到19世纪，它仍然是新古典主义的标志性著作。

本书作者首次精确测量了雅典的古希腊建筑遗迹。书中有多幅建筑物、雕塑、花瓶等装饰物的图像和平面图。《雅典古迹》指出了希腊建筑和罗马建筑之间的差异，改变了人们对希腊建筑的理解，并从根本上挑战了人们关于古典理想的主流观念，引发了此后100多年间欧洲国家和美国建筑设计领域中的希腊复兴风潮。

今据1858年版影印，以飨读者。

THE

ANTIQUITIES OF ATHENS

AND OTHER

MONUMENTS OF GREECE

AS MEASURED AND DELINEATED BY

JAMES STUART, F.R.S. F.S.A., AND NICHOLAS REVETT,

PAINTERS AND ARCHITECTS.

SEVENTY-ONE PLATES.

THIRD EDITION, WITH ADDITIONS.

LONDON:

HENRY G. BOHN, YORK STREET, COVENT GARDEN.

MDCCCLVIII.

LONDON:
Printed by G. Barclay, Castle St. Leicester Sq.

PREFACE TO THE SECOND EDITION.

WE have not been disappointed in our expectation that this little work would make its way. Independently of all other considerations, it appeared that some such manual was really wanted, not only for popular information, but for ready and convenient reference, in circumstances and situations where the voluminous or costly publications which relate to the general subject must be altogether unattainable. A steadily increasing sale has justified our anticipations, and at the same time enabled us in a new edition to improve on the execution of the former. The anxiety to compress had led to the exclusion of matter not merely interesting but instructive, and a somewhat extended reference to original authorities has supplied materials which were before, either not easily accessible, or passed over in the uncertainties of a first experiment. Few corrections have been found necessary, but considerable additions have been made throughout: greater scope has been given to description and explanation: where deficiencies have been detected, the requisite details are supplied:—it has, in brief, been our endeavour to render the body of information ample and complete for all usual purposes, whether practical or popular.

LIST OF PLATES.

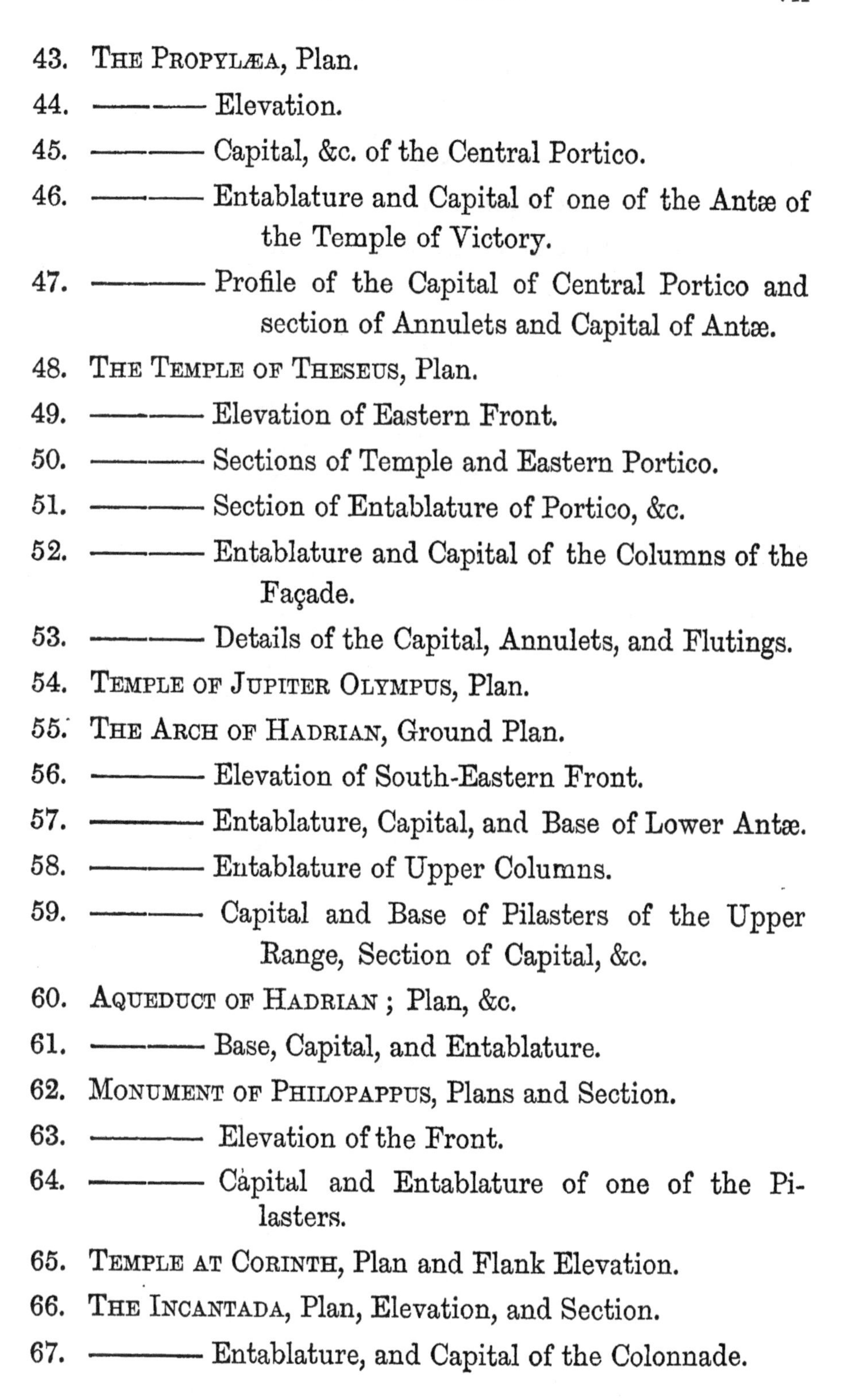

THE

ANTIQUITIES OF ATHENS.

INTRODUCTION.

WITH one paramount exception, Athens is peerless among the existing monuments of the ancient civilised world. The ruins of Rome may be more gorgeous; of Babylon, more mysterious; of Persepolis, more romantic; of the Egyptian Thebes, more vast; but in all that is interesting to thought and feeling,—in memories and associations, deep, affecting, sublime, Athens transcends them all. The beauty of her landscape—the brightness of her sky—her olive-groves—her mountains,—

"The gulf, the rock of Salamis;"—

these still adorn that famous plain of which the Athenian Acropolis is the centre and the crown. Yet these are not the attractions which bring the scholar and the artist to these fair regions: lovely as the face of Nature is, there are still more impressive objects of admiration, in the glorious wreck of those master-works which have made the City of Minerva the wonder of all ages.

It is most unfortunate that the annals of Art should so rarely commemorate its encouragement and conservation, and so commonly tell the story of neglect and destruction. The history of Athens, where Art was carried forward by men such as the world has not since owned, from the simple and severe forms of its earlier efforts to the perfect combination of highest and purest elements, shows no exception to the universal rule. The two great devastators, time and the conqueror, have done their work upon the noblest monuments of human genius; yet those unrivalled constructions which, in their finished beauty and grandeur, were the pride and marvel of antiquity, might have remained nearly entire, but for the persevering encroachment of domestic and dilettante depredation. The accidents of war, the conflagration, and the explosion, may have, once in a century, shattered a temple; but the dilapidator has never ceased. From Sylla who exported an entire colonnade, and Mummius who sent off by ship-loads the treasures of Grecian Art, to the Voivode who worked friezes and inscriptions into the walls of the citadel or of his own house, the mason who breaks down some exquisitely-carved marble into cement, and the amateur who purchases or purloins whatever may be of convenient portage,—from the wholesale to the retail plunderer, all have made the curious and the precious things of Greece their prey. Nor, so far as the virtuosi are concerned, can we regret or condemn the theft, for they have removed their spoil from situations where it was exposed to neglect or mutilation; they have brought within our reach models of an excellence until then unknown; they have contributed essentially to the increase of knowledge, and to the refinement of the public taste.

The course of spoliation would furnish the subject of an interesting dissertation: the limits of this Introduction allow of but a passing reference. The Persian invader consigned to indiscriminate destruction all that was not given up to him with the most abject submission. Cities and temples, villages and olive-groves, were burnt or demolished. But these ravages were repaired: few and obscure are the indications that can now be traced of these remoter visitations. The domestic wars of Greece, wretched and suicidal as they were, left the sanctuaries of the gods and the consecrated works of Art untouched. The Roman conquest swept from both European and Asiatic Greece many of the finest productions of the sculptor and the painter, yet, with few exceptions, there was little wanton damage inflicted; and, in later times, the Cæsars seem to have taken a pride in extending their patronage to their Helladic dependencies. Caligula and Nero were exceptions to this humane and enlightened character; the first from insane vanity, the latter from something like taste, made large demands on the remaining treasures of Art. The iconoclastic propensities of some of the Christian emperors made fearful havoc among temples and idols; nor can the formidable inroads of Alaric the Goth and Genseric the Vandal be omitted, even in this hasty recapitulation of the disasters by which Greece was irretrievably despoiled of her best possessions. The Crusades were another source of evil; and the Turkish conquest completed what the long train of previous inflictions had left undone.

After this brief exhibition of the disastrous fate which has befallen the productions of the great artists of Greece, instead of marvelling that so few have been preserved to the present time, it may well excite our special wonder that so many have survived the casualties to which they have been exposed. It is hardly less a subject of astonishment, that it should have been reserved for the inquiries of so late a period to obtain anything approaching to an accurate knowledge of the state of Greece, with respect to its remaining antiquities. From time to time there had been partial efforts in this way made by individuals, but they seem to have terminated in nothing that was satisfactory, until the latter part of the seventeenth century, when several travellers, mostly invested with a diplomatic character, visited Athens; and the nearly contemporary travels of Dr. Spon and Sir George Wheler first gave authentic, though incomplete, information on the actual condition of Athens and its ancient structures.

It was not, however, until the middle of the succeeding century, that a clear, comprehensive, and scientific survey of these glorious remains of classical antiquity, was made by observers thoroughly qualified by study and practice for an enterprise so bold and arduous. It is to James Stuart that the world is indebted for the first survey, conducted upon scientific principles, of the architectural antiquities of Greece. It occurred to that eminent man, while engaged at Rome in the pursuit of his professional studies, that he might prosecute them to greater advantage on ground more purely and primarily classic, thus seeking knowledge at the very fountain-head of Art. Having associated in his enterprise—which was in part a commercial speculation, including a scheme of extensive publication—his fellow-student, Nicholas Revett, they proceeded to Athens in 1751; and their residence there included a term of nearly three years, during which they were indefatigably employed in exploring, measuring, and drawing, the magnificent ruins by which they were surrounded. In 1761 the first volume of their labours was given to the world, and a new impulse was communicated to the study of ancient Art. The Dilettanti Society soon formed itself, and commenced a series of spirited and skilfully-conducted researches, which have completed and extended the investigations of Stuart. Further down than this it is unnecessary to continue these details: the important additions which have since been made by travellers, both amateurs and professional men, have not indeed exhausted inquiry, but have materially enlarged the circle of its enlightened and successful prosecution.

It would be foreign to the intent and scope of the present manual, were this deduction of facts extended to the wide range of architectural history; but as an introduction to the following pages, it may be expedient to give a brief exposition of their object, in order to a perfect comprehension of the plan. The plates are of French workmanship, from the graver of artists long practised in this sort of reduction; and the selection has been carefully made from the great work of Stuart, so as to include the largest possible amount of instruction and exemplification. Elevations, plans, sections, details, are given profusely in a clear and expressive style of execution. This small volume includes no less than seventy-one plates, exhibiting an extensive illustration of the Greek Orders in the majestic simplicity of their earlier design, the pure and pervading beauty which distinguished their progress and maturity, and the richness which marked even their degradation by the Roman school. Examples are given of the Doric order, from the heavy masses of the Temple at Corinth, to the perfect proportions of the Parthenon; of the Ionic, from the simple but admirable forms of the Temple on the Ilissus, to the exquisite enrichments of the Erechtheum; of the Corinthian, from the graceful luxuriance of the Monument of Lysicrates, to the denser but more commonplace foliage of the Incantada. Much scientific detail and correct measurement will also be found in the plans and sections.

In the explanatory part of the volume, it has been

the anxious endeavour of the Editor to communicate as much information as might be compressible within the limits of a hand-book. Is it too presuming to hope, that in its present form this small but comprehensive manual may be found to supply a real deficiency; that it may furnish the student with a clear and intelligible introduction,—the man of letters with a well-arranged and fully-exemplified system, easy of recollection and reference,—and even the professor with a pleasant and convenient vade-mecum?

THE ACROPOLIS.

PLATES I. II.

IN the greater number of instances the site of the ancient cities of Greece appears to have been determined rather by the position of some insulated rock, of which the platform might be surrounded with a strong and uninterrupted wall, than by the usual circumstances of domestic or commercial accommodation. Even when, as in Athens, the neighbourhood of a commodious haven may have formed one strong inducement to the choice of a particular locality, the settlers rejected the obvious expediency of occupying the shore, for the greater security of some rugged elevation, though at an inconvenient distance. Nothing can illustrate more expressively than this simple fact, the unsettled and insecure condition of Greece in the earlier times. Nor, in truth, did the necessity for similar precautions ever wholly cease: that miserable struggle for supremacy, which kept the independent states in constant agitation, multiplied fortresses in every direction; and the effect of this has been so far fortunate, that many relics of antiquity were preserved from the destruction which marked the course of invading armies or marauding bands.

The Athenian Acropolis was fortified at a very early period; and historical tradition ascribes the construction of its defences to the Pelasgi, that mysterious race, who seem to have been the great masters of military architecture in those ancient and uncertain ages. Thus secured against assault, it became a consecrated precinct, filled with temples, and absolutely crowded with the noblest productions of art. The account given by Pausanias of its sacred buildings and commemorative statues, is "a thing to wonder at!" and the reader is tempted to ask if it were possible that so much of beauty and magnificence could be accumulated within so limited a space. The temples of Diana, Venus, and Minerva Polias, are, with the Parthenon and the "Temple of the Genius of Pious Men,"* mentioned by Pausanias; and it is highly probable, from other authorities, that his enumeration does not include the whole. Of all this glorious show, nothing now remains but the Parthenon, the Erechtheum, and the Propylæa,—shattered, indeed, and deplorably mutilated, but retaining enough of their original form

* This is Colonel Leake's rendering of Σπουδαιων Δαιμων: it can, however, hardly be taken as a very satisfactory interpretation. Taylor gives it, "the Demon of Worthy Men;" and Clavier evades the difficulty by keeping close to the original, "*le Genie Spoudæon.*" Schubart and Walz, in the most recent critical edition of Pausanias, adopt the average interpretation, "*bonorum virorum genio.*"

to prove how much of the beautiful and sublime they must have exhibited in their perfect state.

Yet there is one circumstance, of comparatively recent discovery, and still more recently ascertained to its full extent, which gives a strange contradiction to our cherished notions concerning the purity of Grecian taste, and its antipathy to all coarseness and exaggeration. It should seem that the Greeks *painted their temples,* not merely in chiaroscuro, or in subdued tints, for the purpose of giving relief to projections or expressiveness to ornamental details, but with glaring colours,—reds, and blues, and yellows; with violent contrasts, the columns one hue and the entablature another. Nay, there is shrewd suspicion that the sculptures were painted, like the figure-head of a man-of-war, and that the pillars were *striped!*—the flutings being left of the unstained marble, while the rest was daubed with villanous ochre. And, unluckily, the evidence for these incredibilities is most exasperatingly clear; the statements of the German architects employed by King Otho, and the very interesting details given by Mr. Bracebridge, leave no doubt whatever of the facts. Still, we have our doubts; not, indeed, as to the correctness of the testimony, but respecting the date of the practice. We cannot believe that the architects of the best days of Greece would so carefully select the purest materials in the prospect of their concealment by a mask of tawdry colour,—that they would give such an anxious finish to their carvings, knowing that their sharpness and delicacy would be impaired by the brush of the "ornamental painter." Neither is it probable that, if this vile practice had existed in the olden time, no hint of it should occur in Pausanias or Vitruvius. That the Greeks used colour on the exterior of their temples—at least, that there are now to be found upon them traces of colour, cannot be questioned; but that Ictinus and Callimachus, to say nothing of Phidias and Praxiteles, practised these atrocities, while Pericles approved and patronised, can only be believed—*quia impossibile est.*

In the first edition, nothing more than this was said in illustration of the Acropolis, considered as a repository, surpassingly rich, of the noblest productions of Art. It seems, however, to offer but a slight and insufficient notice of objects which would require volumes to describe fairly; and it may tend, in some degree, to supply this obvious deficiency, if we extract from the "Attica" of Pausanias, a few additional indications of that glorious scene as it existed in his day. Deeply as we are indebted to this active and observant traveller for the information which he has left us on almost all subjects connected with Greek antiquity, it is impossible not to regret that he should have given it in so incomplete and disjointed a form. We have in the "Periegesis" hints of the greatest importance, but of the most provoking brevity; scraps and shreds without coherence or sequence; facts without the necessary comment, and comments with imperfect facts. In one place, he withholds invaluable explanations because he has been warned to silence by a dream; in another, he contents himself with a simple assurance that he is perfectly well informed on the subject, but that he will not communicate. Even where no scruple, no prohibition can possibly intervene, he contents himself with a mere repetition of his travelling memoranda, and thus leaves many an interesting point in hopeless uncertainty. Yet we may well hold his memory in grateful admiration for that which he has preserved; and, perhaps, the very form and quality of his communications may have aided in their conservation. A larger work would have been less frequently copied, and with more difficulty kept from injury; nor ought it to be overlooked, in our regret that such ample materials have not been transmitted to us in a more comprehensive and compact form, that after all it must have been impossible to give anything beyond the mere outline of a subject so vast. The History, the Mythology, the Topography of Greece, with all that was incident to these in fact and fiction, and all that might illustrate them in existing institutions, public monuments, or living manners, lay before him to observe and to describe: few men would have been equal to such a task,—Pausanias most assuredly was not.

Such, however, is our best, indeed our only direct authority for the details of the Acropolis; and the accuracy, as well as the incompleteness of his description, is sufficiently attested by the actual remains. He commences with the triple port of Athens, and after very summarily noticing the more striking objects that presented themselves on the two roads leading to the city from Piræus and Phalerum, he makes an irregular circuit before he enters the sacred enclosure of the citadel. Like every other part of this marvellous construction, the wall itself might serve as the text of a lengthened, yet interesting disquisition. It was built or restored at different periods: the earliest portion was on the northern line, and tradition assigned its execution to the Pelasgi, though it would seem that some of the peculiar forms of those mysterious architects are not now to be observed among the remains. Cimon, the son of Miltiades, was the reputed builder of the southern wall: this, however, is questioned by Dodwell, who supposes that the Κιμωνιον τειχος was an interior fortification, and that the entire peribolus was constructed by the Pelasgi. Among his authorities for this most gratuitous supposition, he refers to Pausanias, erroneously as it should seem, since, if we may trust our own examination, that writer distinctly *excepts* the part built under the superintendence of Cimon.

In the Arts, as in all else that related to intellectual enjoyment, the Greeks turned everything to account. The rock on which the Athenian Fortress stood in all its pride of ornament and strength, was itself made either the material or the mould of many a graceful structure. Its recesses were consecrated either by historical or religious associations, and decorated with skilful adaptation to their especial character: the irregularities of its circuit were shaped into theatres for music and pantomime: columns, statues, and tripods, occupied its minor projections. The wall itself of the superior platform was, in various ways, charged with adornment,—in some instances, not altogether in harmony with the normal principle which inseparably connects utility with beauty. To the face of the southern wall was attached an Ægis, bearing on its centre the head of Medusa, gilt. On the eastern end of the same line stood several groups of statuary: the war of the Giants, the battle of the Amazons and the Athenians, the fight of Marathon, and the extermination of the Gauls in Mysia. Of these, the dedication is ascribed to Attalus. The height of the figures (if, indeed, the words relate exclusively to height) appears not to have exceeded three feet. Nothing is said of the execution of these statues, nor of their effect when contemplated from below: it seems, however, difficult to imagine that it can have been good in any position. Viewed from without, they must have looked like puppets,—exciting no small wonder how and why they were thus ranged, in a situation so awkwardly exposed to every casualty of weather, accident, and mischief. It is, indeed, recorded, that the statue of Bacchus, connected with the Gigantomachia, was blown from its "pride of place," during a violent tempest.

To the *terre-plein* of the rampart there is but one approach, and that in the days of old was adequately fortified against the modes of assault then in practice. Of the system adopted in constructing the immediate entrance to the Acropolis, the details will be found under the proper head; and nothing further will be said in this place of the Propylæa, than to suggest, that, when the architecture of Greece is censured for restricted range, this most impressive group of buildings may be referred to, in evidence that the Greeks themselves found no difficulty in adapting its elements to every purpose of useful and ornamental structure.

Passing onward through this unrivalled avenue, and standing beneath its eastern portico, the eye of the traveller would command the entire perspective of the Acropolis, and in the distance the great mountain-ranges of Attica. On the right stood the Parthenon, majestic in its height, and beautiful in its exact proportions. To the left and in front were the Erechtheum, the Cecropium, and other structures of sacred character; while the intervals and vacant spaces of the enclosure were filled up by statues, altars, and other memorials of Gods, Heroes, Patriots, and Bards. Of all these nothing but the wreck remains,—the relics of an age and people such as the world has not since possessed.

Plates 1 and 2 are representations of the Athenian Acropolis in a restored state. The first exhibits the eastern aspect, where was the only approach to the platform; and in this view the letters A, B, C, indicate in succession the Propylæa, one of the flanking temples, and the Parthenon. The second shows the northern side, and the same letters refer to the same structures; D points out the Erechtheum. Of all these structures, ample illustrations are given in the following pages, which contain, in addition to an extensive collation of authorities, extracts, analytical and descriptive, from architects and travellers, wherever it was thought that the language of actual observation might give distinctness to complicated details, or interest to an impressive object.

DORIC PORTICO,

OR,

GATE OF THE AGORA.

PLATES III. IV.

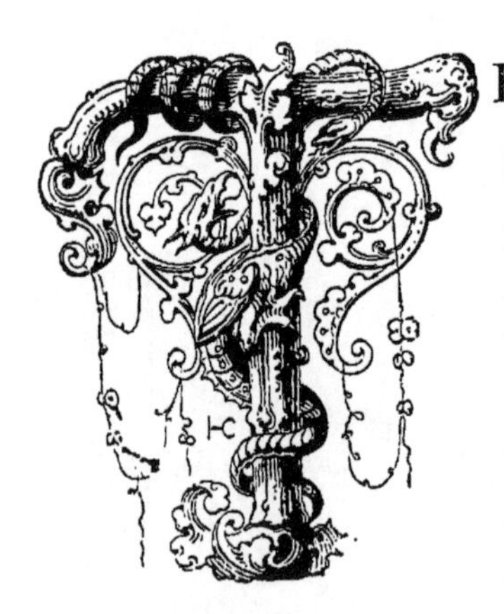

HE ruin of which the third and fourth plates give the authentic restoration and the geometrical proportions, had long been considered as the only remaining fragment of a temple dedicated to Rome and Augustus. Stuart, however, on grounds far more satisfactory than those which had given currency to this belief, suggested that these columns might once have flanked the entrance to a public market; and all subsequent research has tended to confirm his opinion. In addition to peculiarities of construction which have not been found in the remains of sacred edifices, the testimony of inscriptions is decisive. Of these there are several connected with the building; and one of them records the names of two "Prefects of the Market," while another preserves an edict of the Emperor Hadrian, regulating the sale of oils, and the duties payable on that common article of traffic. There were in Athens, two of these structures; one of them, probably that to which this portico belonged, was distinguished as the New Agora; it was founded by the Cæsars, Julius and Augustus, and among its decorations were statues of the Cæsarean family.

PLATE III.

Fig. 1. Plan of the Portico,—on the jamb, to the left of the spectator, marked A, is the inscription containing Hadrian's decree. The prolongation of the wall connected with these jambs, beyond the side walls B, is contrary to all usage in the construction of temples, and evidently indicates a structure intended for other purposes.

Fig. 2. Geometrical elevation of the front of the Portico.—The Acroterium on the summit of the pediment probably supported an equestrian statue of Lucius Cæsar.

PLATE IV.

Fig. 1. Details of the capital and entablature.

There is a slight but marked distinction between the proportions of these columns, and of those which the Greeks attached to sacred edifices, in conformity with the principle affirmed by Vitruvius in the recommendation that in buildings of secular character, the proportions should be less massive and imposing than those which were employed in the construction of temples. It is suggested by the editor of Stuart, that "this example of the Doric order, authenticated to be the latest of a purely Grecian character, affords a greater facility of adaptation to the modern practice of domestic architecture, than any antique model extant."

The Greeks appear, indeed, to have been far from solicitous for the strict maintenance of mere symmetry, when important objects required its sacrifice. Whereever it was practicable to combine uniformity with use and ornament, the Grecian artists adhered to it as one of the surest elements of architectural effect; but when circumstances demanded a relaxation of the rule, they seem to have felt neither scruple nor difficulty in availing themselves of the ample resources supplied by their genius and skill. Of this there is clear illustration in the Erechtheum and the Propylæa. It would, probably, have been easy in both instances to maintain, if not perfect regularity, something much more nearly approximating to it than we find to have been the case. The obstacles interposed by the irregularity and abruptness of the rocky surface might have been overcome by levelling or substruction; and with respect to merely technical difficulties, the resources of art were fully equal to their removal. Local superstitions, however, were a more unmanageable matter, and to their influence must be mainly ascribed the departures from symmetrical arrangement which occasionally present themselves in the works of the Greek architect.

Fig. 2. Profile of the capital on a larger scale.

IONIC TEMPLE ON THE ILISSUS.

PLATES V. VI. VII.

NEAR the fountain of Callirrhoe, on the southern bank of the Ilissus, stands a small temple, of the Ionic order, but differing considerably in its details from all ordinary example. The forms are simple but elegant, and the execution is in all respects so perfect that this building may be considered as one of the most remarkable productions of Grecian architecture.

At a period now uncertain, this temple was repaired after a barbarous fashion, and converted into a church, sacred to "Our Lady of the Rock," a name which it retained in the time of Stuart, although deserted and miserably shattered. Since then it has been entirely destroyed. Much doubt exists respecting the original dedication. Spon assigned it to Ceres; Stuart, to the hero Panops; but Colonel Leake and Sir William Gell, with greater probability, suppose that it was the Temple of Triptolemus.

PLATE V.

Fig. 1. Plan of the Temple, amphiprostyle.—A, the portico; B, the pronaos; C, the naos, or cella of the temple; D, the posticum; E, one of the antæ. The two middle columns of the portico no longer exist; but, on the place where they ought to stand, there are traces of circles, exactly equal in diameter to the remaining columns; and these indications sufficiently prove the intention of the architect.

At the time of Stuart's visit, the capitals of the antæ of the posticum were in excellent preservation; in width they were of the same dimensions with those of the portico, but they had only half the thickness; while in those of the pronaos, the faces E, F, were equal.

Fig. 2. Elevation of the Portico in its perfect state. —It is conjectured that the frieze was ornamented with bas-reliefs.

PLATE VI.

Details of the Base, Capital, and Entablature, as restored by Stuart.—The figures on the frieze are copied from a fragment found at Athens, and which so exactly agreed in dimensions with the place assigned to it, that there is no improbability in supposing them to have been originally in combination.

The columns of this temple exhibit peculiarities which have led to the suggestion that it may have been one of the earliest specimens of the Ionic. The shafts are "shorter" and "less diminished;" the capitals are larger in proportion to the entablature, than occurs in later works: the bases, too, are without plinths. It has, moreover, been observed, that in the details of the base, there are marked resemblances between these columns and those of the colossal temple of Juno, at Samos, probably the most ancient Ionic structure of which the remains have been explored in modern times.

PLATE VII.

Capital and Base of one of the Antæ; with representations of the different architraves which belonged to the several parts of the temple.—The mouldings of the capital and the base are carried round the exterior of the building; but in the interior of the pronaos, the base only is continued.

Fig. 1. Architrave of the portico.

Fig. 2. Architrave of the pronaos.—The upper fascia was enriched by a painted ornament.

Fig. 3. Architrave of the posticum.

It is deeply to be regretted that of this interesting structure, so few particulars should have been preserved. Its simplicity and marvellous beauty made it a sort of test of excellence in Art; and were it only for what he has done in regard to this fine memorial of Grecian taste, Stuart's name would deserve grateful remembrance: having done so much, it were unreasonable to require what he was probably prevented by circumstances from effecting. The hope of further illustration has, however, been altogether destroyed by a series of disastrous events. Having been converted to the uses of the Greek Church, this would have insured its preservation; but in 1674, a French ambassador took it into his head to celebrate mass within its precincts, and this desecration caused it to be abandoned by the Greeks. Thus neglected it became ruinous, and in 1780 was demolished by order of the Turkish Voivode, and the materials used in building.

OCTAGONAL TOWER

OF

ANDRONICUS CYRRHESTES.

PLATES VIII. IX. X. XI.

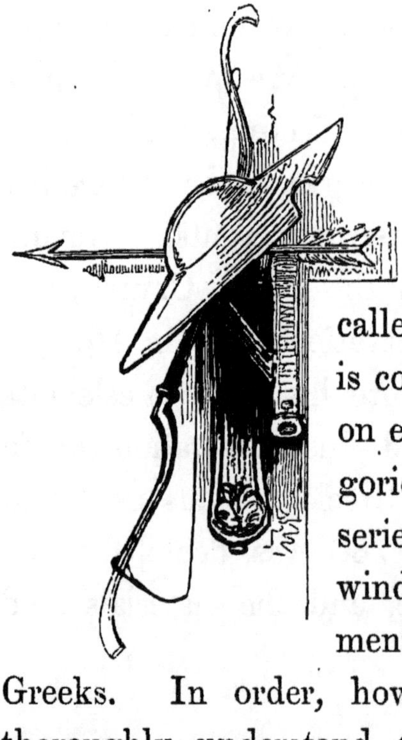

THIS singular building, usually called "The Tower of the Winds," is constructed of marble, and bears on each of its eight faces an allegorical figure in relief; the entire series representing the different winds, according to the arrangement and nomenclature of the Greeks. In order, however, that the reader may thoroughly understand the character and object of this curious and complicated edifice, it may be expedient to cite the clear description of Vitruvius, as ranslated by Stuart: "Some have chosen to reckon only four winds: the East, blowing from the equinoctial sun-rise; the South, from the noon-day sun; the West, from the equinoctial sun-setting; and the North, from the Polar stars. But those who are more exact, have reckoned eight winds, particularly Andronicus Cyrrhestes, who on this system erected an octagon marble tower at Athens, and on every side of the octagon he wrought a figure in relievo, representing the wind which blows against that side: the top of this tower he finished with a conical marble, on which he placed a brazen Triton, holding a wand in his right hand; this Triton is so contrived that he turns round with the wind, and always stops when he directly faces it; pointing with his wand over the figure of the wind at that time blowing."

This description applies, with the utmost exactness, to "The Tower of the Winds." Four of its faces front the cardinal points: the part immediately below the cornice bears on each of its eight divisions a figure, skilfully designed and wrought, representing the supposed characteristics of the wind to which it was inscribed. Beneath these figures are traced solar dials, to the correctness of which the celebrated Delambre bears testimony, and describes the series as "the most curious existing monument of the practical gnomonics of antiquity." The roof is of marble blocks, wrought into the form of tiles. There are two entrances, facing respectively to the north-east and north-west: each of these openings has a portico, supported by two columns. When Stuart explored this building, the lower part of the interior was covered to a considerable depth by rubbish; and the Dervishes, who had taken possession of the building, performed their religious rites on a wooden platform which had been thrown over the fragments. All this, however, he was permitted to remove, and he found manifest traces of a clepsydra, or water-clock, carefully channelled in the original floor; thus completely illustrating the term Horologium, applied to the tower by Varro, and exhibiting both the simple and the scientific mechanism employed by the ancients for the measurement of time. Further details will occur in the explanation of the plates.

PLATE VIII.

Fig. 1. Plan of the Tower of the Winds.—The attached circular portion, of which the general exterior appearance may be observed in the plate immediately succeeding, probably contained the reservoir which supplied the water to the clepsydra; and the channels marked on the floor were, no doubt, connected with the machinery, though in what way cannot now be ascertained with precision.

Much is wanting to the history of this celebrated Tower. Varro and Vitruvius both describe it, and the probable period of its construction may be referred to the second century before Christ,--when Scipio Nasica set up a similar building at Rome, of which the architectural character does not appear to be now ascertainable. The stream which gave motion to the machinery of the Athenian time-piece had its spring in the cave of Pan on the flank of the Acropolis, and portions of the artificial channel by which it was conveyed, are still to be traced in the intermediate space.

An inspection of the plan and section will show that the building is interiorly ornamented by cornices at different heights, and that their projections are indicated by simple lines, while the substance of the wall is exhibited by a broad and shaded band. The wall varies somewhat in thickness, and the variations are distinctly expressed in the figure, which is a careful reduction from the original engraving. That every advantage may be afforded for the comprehension of these details, we shall here give the references and explanations from Stuart's own text: "Each external face of this octagon tower, considered without its ornaments, is one perpendicular plane from top to bottom; but on the inside it is otherwise; for that part of each face which is above the second cornice, projects two inches over the part which is between the said cornice and the pavement. The lowest of the interior cornices is interrupted by the two doors, and breaks off on each side of them in a very obtuse angle; and the upper cornice or entablature, supported by eight columns, as likewise the fascia on which those columns stand, are circular. So far, therefore, as the plan regards these particulars which are on the inside of the tower, it is necessary to divide it into four parts. The first part, from *a* to *b*, is one-fourth of the interior surface of the wall immediately above the pavement; the second, from *b* to *c*, is one-fourth of the interior surface immediately above the lower cornice; here the greatest projection of this cornice is marked by a single line, and the manner of its breaking on each side of the doorways is also shown: the third part, from *c* to *d*, is the interior surface of the wall above the second cornice; the projection of this cornice is also marked with a single line: the last part, from *d* to *a*, is the remaining fourth of the interior surface; on this is marked the circular band, or fascia, on which the eight columns are placed, with the plans of two of those columns."

Fig. 2 exhibits the entire section of the tower; and an examination of this plate will make still clearer the full explanation given above. The first and second cornices,—the circular fascia, or plinth, supporting the columns, which find room by occupying the angles of the wall,—the entablature, or upper cornice,—and the roof,—are distinctly marked.

PLATE IX.

Elevation of the Tower of the Winds.—This restoration has been made on the authority of the existing remains, and of fragments found on the spot. The Triton is from the description of Vitruvius, and has always appeared to us a singularly awkward contrivance. It would, no doubt, be exceedingly difficult to adapt a graceful form to the purposes and conditions required in the present instance, but the Greek architects were in the habit of overcoming greater embarrassments than these; and it is hard to believe that this strange merman, with his tail in one hand and a stick in the other, can present a single line or feature of the original design.

PLATE X.

Capital and Entablature of one of the Porticoes.—It should be stated that the columns are much defaced, and that these members are delineated from fragments which, although found on the spot, can only be assigned to their respective places on highly probable grounds.

There can, however, be little ground for hesitation in this matter, since the character of the crowning member, on which the "Triton" rested, and of which, though not actually found in its proper position, the authenticity cannot be for a moment questioned, is so completely in harmony with other remaining fragments, as to set the question entirely at rest. The Capital is commonly assigned to the Corinthian order, though deficient in several of its leading features. The Corinthian acanthus leaves rising tier over tier, with their scrolls and other ornamental adjuncts, are represented by a single wreath of enriched foliage, and above it a course of flat and nearly plain water-leaves, Egyptian rather than European in their expression, while the abacus is altogether unadorned. Still, taken without reference to systematic arrangement, this capital is pleasing in form and effect, and from its simplicity readily applicable to domestic architecture. With the Choragic Monument, it has absolutely nothing in common, and when they are placed in connexion as the earliest and latest examples of the Greek Corinthian, the arrangement is wholly at variance both with fact and sound principle.

PLATE XI.

Fig. 1. Profile of the exterior cornice. — The lion's head, which ornaments the cymatium, is perforated, and serves as a gutter to carry off the rain-water.

Fig. 2. A fourth part of the roof,— the cavity in the centre, marked A, was probably the socket of the capital, or ornamental base, which supported the Triton.

Fig. 3. Section of half the roof,—the dotted prolongation of the line of the roof is intended to show how deeply the capital was engaged before this part of the building was damaged.

The circumstances connected with this extraordinary structure have not received sufficient investigation. It might serve as the text to many an important inquiry; and its position, both local and chronological, suggests a series of questions much less easy to answer than to propose. The whole aspect and proportions of the building have no alliance, it would seem, with Greek taste and feeling. The execution of the emblematic figure is, indeed, highly praised by Stuart, but it appears to be essentially Roman, and reminds us far more of the Trajan Column than of the pediment and metopes of the Parthenon. The roof is skilfully constructed; and it has been already stated, on the high authority of Delambre, that the dialling of the different faces displays admirable science. It is much to be desired, that some one of thorough qualification, both as architect and archaiologist, would give his leisure to the investigation of this instructive monument.

THE CHORAGIC MONUMENT OF LYSICRATES.

PLATES XII. XIII. XIV. XV. XVI. XVII.

SERIES of temples forming a street is one of the features of Athens. "These temples were surmounted by finials which supported the tripods gained by victorious Choragi in the neighbouring theatre of Bacchus, and here dedicated by them to that deity, the patron of dramatic representation. Hence the line formed by these temples was called the Street of Tripods. From the inscriptions engraved on the architraves of these temples, recording the names of the victorious parties, and the year in which the victory was gained, the dramatic chronicles, or Didascaliæ, were mainly compiled. Thus these small fabrics served the purposes at the same time of fasti, trophies, and temples."

Of these structures this monument must have been one, and surely the most beautiful. Exquisitely wrought, graceful in its proportions, rich in decoration, it only required for its perfection that the material should not disgrace the design and execution. Fortunately this was close at hand: the fine-grained marble of the Pentelic quarries enabled the Athenian architects, not only to produce the happiest effects, but to maintain throughout that faultless elaboration which distinguishes the purest examples of Grecian art. In this edifice, the roof, the base of the colonnade, and the shafts of the columns, are each of one block. The frieze and architrave are also, unitedly, of one piece, and the masses of stone, which form the steps of the stylobate, are entire. The whole building consists of a quadrangular basement, supporting a circular temple crowned with a tholus, or cupola, terminating with an ornament, on which stood the tripod, of which an inscription recorded the dedication. As there are peculiarities of construction in this edifice which deserve distinct specification, we shall give in explanation the description of Stuart himself, since, in all cases of difficulty or complication, the definitions of an actual observer have the best chance of being intelligible: "The colonnade was constructed in the following manner; six equal panels of white marble, placed contiguous to each other on a circular plan, formed a continued cylindrical wall, which of course was divided from top to bottom into six equal parts, by the junctures of the panels. On the whole length of each juncture was cut a semi-circular groove, in which a Corinthian column was fitted with great exactness, and effectually concealed the junctures of the panels. These columns projected somewhat more than half their diameters from the surface of the cylindrical wall, and the wall entirely closed up the intercolumniation. Over this was placed the entablature and the cupola, in neither of which any aperture was made, so that there was no admission to the inside of this monument, and it was quite dark."

Yet have the ingenious men of Athens been pleased, in modern times at least, to call this dark inclosure the "Lantern of Demosthenes,"—*lucus a non lucendo*— and to suppose that this lantern without light, this six-feet-wide closet without window or entrance, was actually the study of that great statesman.

PLATE XII.

Elevation of the Choragic Monument.—Neither here, nor in the following illustrations, has anything been restored without authority.

The comparatively high state of preservation in which this admirable production of an unrivalled school was found by Stuart, appears to have been the result of circumstances which could not easily have happened elsewhere. There were in Athens, while it yet remained the city of Minerva, a great number of these edifices; so many, in fact, as to form an entire street. Every Choragus, when he retired from office, placed the consecrated tripod, which was given as an honorary reward for his heavy expenditure, in a temple, of small dimensions but costly decoration, inscribed with the name of the founder and the particulars of the contest. Structures like these, beautiful as they might be, and even in proportion to their beauty, were ill-suited to the fierce and vindictive spirit which too often prevailed among the polities of Greece. The lofty and strongly-built constructions which sheltered the worshippers of Zeus and Athene, might withstand the casualties of civil broil, or the more deliberate injuries of an invading enemy, but the slender and fragile members of these gems of architecture could offer but weak resistance, and have yielded altogether to the various forms of spoliation or destruction by which they were assailed. Two only remain, and these owe their preservation to the peculiar circumstances of their position. The memorial of Thrasyllus is attached to a cave in the solid rock, and its partial security is due to the occupation of the recess by a shrine of the Panaghia. The monument of Lysicrates has escaped destruction from a different cause: it is built up in the wall of a Capuchin convent or *hospice,* where it seems to have served the various purposes of closet, oratory, and library. For this, as for other services to the cause of art and learning, gratitude is due to the monastic brotherhoods, and while arrested in admiration before this faultless relic of Attic genius, let it not be forgotten that we owe its conservation to the good taste and right feeling of a Franciscan recluse.

PLATE XIII.

Fig. 1. Plan of the Monument.—The parts more darkly shaded show the existing portions of the edifice, including three panels and the whole of the columns. The columns are fluted only on the exterior half-circle; the inner semi-diameters are less in radius by half an inch.

Fig. 2. Section of the Monument.—In the interior, the capitals are only blocked out.

Fig. 3. Profiles of the base of the columns, and the cornice of the basement.

PLATE XIV.

The Entablature; the exterior face of the capital; and the half of one of the tripods, which are wrought in relief on the upper part of the intercolumniations, immediately below the architrave.—Although none of the capitals were complete, a careful collation enabled Stuart to give the whole with minute accuracy: so scrupulous was he in this respect that, as may be seen in the following plate, he abstained from the restoration of a fragment of foliage, not having been able to trace with entire exactness the original form. The annular channel between the shaft and the capital is supposed to have contained an astragal, or collarino, of bronze. The small figure on the left represents the profile of the fascia and moulding, below the tripods.

PLATE XV.

Fig. 1. The "flower," or crowning ornament, on the top of the tholus.—The letter A, at the side, refers to an arrangement of foliage so much injured as to baffle all attempts at restoration.

Fig. 2. Plan of the upper surface of the "flower." —A, A, A, cavities formed to retain the feet of the tripod. B, socket of the central support of the tripod.

Fig. 3. L, perpendicular section of so much of the upper part of the flower, as may serve to show the depth of the cavities at A and B, in the preceding figure.

PLATE XVI.

Fig. 1. A fourth part of the upper surface of the roof.—This beautiful exterior is worked with great delicacy in the form of a sort of thatch of laurel leaves, surrounded by an ornamental edge, usually termed a Vitruvian scroll. A, one of the three helices, caulicŏli, or scrolls, which form the triple division of the roof. B, a cavity which probably held some bronze ornament.

Fig. 2. Section on the line C D, of fig. 1.

Fig. 3. Partial section of the copula on the line E F, showing the arrangement of the leaves.

Fig. 4. Section of the helix, or scroll, marked A, in fig. 1. This section is on the line *a b*, fig. 2.

Fig. 5. Examples of the Vitruvian scroll, which surrounds the tholus; and of the antefixæ, which ornament the cornice.

PLATE XVII.

Fig. 1. Plan of the Capital.—The segments C D, represent the interior half. The letters E, F, G, H, refer to the other division. C is an horizontal section at the line indicated by the same letter, fig. 2. D, a similar section at D of the same figure. E, F, G, exhibit different plans, expressing sections marked by corresponding letters attached to fig. 4.

Fig. 2. Elevation of half the internal unfinished face of the capital.

Fig. 3. Vertical section through the axis of the unfinished inner half.

Fig. 4. Vertical section through the centre of the finished exterior half.

The frieze is ornamented by figures, of which a specimen is given in Plate XIV. representing the punishment of the Tyrrhenian pirates by command of Bacchus. Nothing can be more spirited than the execution of these groups, nor more expressive than the way in which the story is told.

The beautiful modification of the Corinthian capital, which distinguishes this monument, has been supposed to be the earliest known example of that order. This, however, is in some degree doubtful; though there can be no question of its great antiquity.

The profusion with which embellishment was lavished on this beautiful structure, is nowhere more remarkable than in the highly ornamented junction of the shaft with the capital. Instead of the common termination of the fluting, it is finished off into a sort of leaf; while the interval between the annular channel and the proper commencement of the capital is filled up with a circlet of simple but graceful foliage.

PANTHEON OF HADRIAN.

PLATES XVIII. XIX. XX. XXI. XXII.

THIS wreck of the magnificent structure passed among the modern Athenians as the palace either of Pericles or of Themistocles, a guess only resorted to in the utter ignorance of genuine tradition and architectural appropriation. Wheler and Spon supposed these ruins to have belonged to the temple of Jupiter Olympius; an obvious error, since the ascertained remains of that gorgeous edifice occupy a different locality. Stuart, by an exceedingly ingenious deduction, made it appear highly probable that the Poikile Stoa, or painted portico, occupied this site; and he accounted for certain incongruities in the architecture by the supposition that extensive repairs had interfered with the original design. Dodwell adopted the same opinion. Dr. E. D. Clarke thought that these remains might have belonged to the Old

Forum of the Inner Cerameicus. Mr. Wilkins appears to have been the first to suggest that on this extensive platform stood the *Hieron,* or "Sanctuary common to all the Gods," built by the orders of Hadrian; and this opinion has the sanction of the late Sir William Gell. The most complete exposition, however, of the facts connected with this difficult inquiry, is to be found in the notes appended by Mr. Kinnaird to the second edition of Stuart and Revett's great work; and it seems to be there fairly shown, from the language—though not quite free from obscurity—of Pausanias, and from the results of recent examination, that these shattered walls and broken columns formed part, as maintained by Wilkins and Gell, of the splendid Pantheon of Hadrian.

These ruins were, to cite the description of Mr. Wilkins (*Atheniensia*), "the peribolus of a sacred building. The walls next the street are adorned with Corinthian columns advanced before them: in the centre is a portico of four columns, through which the area within is approached. The line of the walls is interrupted by several projections forming cellæ, or chapels, some circular, and some rectangular. Around the walls within was a cloister, or portico, formed by a continued row of columns twenty-three feet distant from them."

All this sufficiently agrees with the description of Pausanias, who speaks of it as "most admirable," with its hundred and twenty columns of Phrygian marble;

its splendid halls with ceilings and ornaments of gold and alabaster; its pictures and statues; and, better even than these, its library. A singularly fortunate verification of its identity was obtained, long subsequent to the visit of Stuart, through the interference of the Earl of Guildford. That enlightened and munificent nobleman obtained permission to excavate within the enclosure, although it had been appropriated to the domestic uses of a Turkish officer; and this decisive experiment at once ascertained the local identity by a discovery of the very "Phrygian stone" described by the ancient traveller. Before these interesting ruins could be reached, it was necessary to remove an accumulation of soil and rubbish covering the original level to a depth of thirty feet.

This structure, however, with all its richness, is characterised by the degraded taste of the Roman architecture. The proportions depart from those of the pure Greek models: the columns are raised on pedestals: the details of the Corinthian architrave are altered for the worse; and that unerring sign of degraded taste, the broken entablature, everywhere prevails.

PLATE XVIII.

Fig. 1. Plan of the Pantheon.—The parts still remaining are shaded; the restorations are merely traced. The walls indicated in the centre mark ancient foundations, on which has been built a church dedicated to the Panaghia.

Figs. 2 and 3. Parts of the front, drawn to a larger scale, to show more accurately the construction of the side walls.

PLATE XIX.

Fig. 1. Half of the general elevation, exhibiting the portal and one division of the front.—At the termination on the left hand may be seen one of the circular projections, exhedræ, cellæ, or chapels, by each of which names they have been designated, as one or other theory prevailed. Nearly the whole of this restoration is sanctioned by existing masses, or fragments. It deserves notice that the abacus of the capital is continued throughout beneath the architrave.

Fig. 2. Section of the front wall, showing the profile of the portal, and of the southern pteroma, with one of the columns which stand between the portal and the northern pteroma.

Fig. 3. Section of the portal, and of the entrance before which it stands.—The interior varies from the exterior architrave.

Fig. 4. Part of the external face of a lateral wall.

PLATE XX.

Capital and entablature of the columns of the front.—The angle of the abacus is acute, like that of the Temple of Vesta, at Rome.

PLATE XXI.

Profiles of the base and pediment of the columns.

PLATE XXII.

Fig. 1. Plan of the capital.

Fig. 2. Angular view of the capital.

These details are altogether insufficient for practical purposes, and leave much to be desired even for general illustration. It does not, indeed, appear that, surrounded as it is by structures of purer taste and more easy access, this splendid monument of Roman magnificence has attracted in a due degree the attention of architectural travellers. Stuart's description is exceedingly meagre, and, under the circumstances, it was hardly possible for him to make it more complete; it may, however, be hoped that more assiduous examination under better auspices will enable modern explorers to supply satisfactory information respecting an edifice of which the character and arrangements are still so imperfectly understood.

THE PARTHENON.

PLATES XXIII. XXIV. XXV. XXVI. XXVII. XXVIII.

F this glorious edifice, built under the auspices of Pericles, Phidias was the designer; Callicrates and Ictinus were the architects. Sir William Gell says: "It is, without exception, the most magnificent ruin in the world, both for execution and design. Though an entire museum has been transported to England from the spoils of this temple, it still remains without a rival."

Respecting the arrangement of this marvellous temple, in its original state, there exists, even among those best qualified to judge, considerable difference of opinion. It seems to be agreed that Stuart relied too much upon the authority of Wheler and Spon, who, indeed, saw it when in a condition of much greater completeness than it exhibited at the date of his visit, but with many inferior advantages in point of architectural science. Subsequently to the investigations of Stuart and Revett, these majestic ruins have been subjected to strict examination by individuals from whose decision there can be no appeal, and among these Mr. Cockerell stands eminently distinguished. Colonel Leake, in his "Topography of Athens," expressly refers to this gentleman as an ultimate authority in cases of doubt; and since the statements of the Colonel have thus the double sanction of his personal observation, and the results of Mr. C.'s more recent and minute exploration, they shall be given in his own words:—

"The Parthenon, or great Temple of Minerva, stood upon the highest platform of the Acropolis, which was so far elevated above its western entrance, that the pavement of the peristyle of the Parthenon was upon the same level as the capitals of the columns of the eastern portico of the Propylæa. The Parthenon was constructed entirely of white marble from Mount Pentelicum. It consisted of a cell, surrounded with a peristyle, which had eight Doric columns in the fronts, and seventeen in the sides. These forty-six columns were six feet two inches in diameter at the base, and thirty-four feet in height, standing upon a pavement, to which there was an ascent of three steps. The total height of the temple above its platform was about sixty-five feet. Within the peristyle, at either end, there was an interior range of six columns, of five feet and a half in diameter, standing before the end of the cell, and forming a vestibule to its door; there was an ascent of two steps into these vestibules from the peristyle. The cell, which was sixty-two feet and a half broad within, was divided into two unequal chambers, of which the western was forty-three feet ten inches long, and the eastern ninety-eight feet seven inches. The ceiling of the former was supported by four columns, of about four feet in diameter, and that of the latter by sixteen columns, of about three feet. It is not known of what order were the interior columns of either chamber. Those of the western having been thirty-six feet in height, their proportion must have been nearly the same as that of the Ionic columns of the vestibule of the Propylæa; whence it seems highly probable that the same order was used in the interior of both these contemporary buildings. In the eastern chamber of the Parthenon the smallness of the diameter of the columns leaves little doubt that there was an upper range, as in the temples of Pæstum and Ægina."*

* In addition to these valuable details, we shall take the liberty of inserting a highly interesting extract from Mr. Kinnaird's elaborate comment: "Commenced about the eighty-third Olympiad, or about 448 B.C., the rapidity of the execution of this fabric is recorded by the historian; and by the comparison of historic dates and events, sixteen years is the utmost extent of time that can be possibly supposed to have been occupied in the performance of the entire works of this edifice, 101 feet in front, 227 in length, and 65 in height, wrought in the most durable marble, and with the exquisite finish of a cameo; enshrining the chryselephantine colossus with all its gorgeous adjuncts,

It is a most unfortunate circumstance that Pausanias, the main authority for all that relates to the great monuments of Grecian art, completely deserts the inquirer at this important point. He despatches the Parthenon in brief and insignificant phrase, so far as the building and its peculiarities are concerned, while he describes, in considerable detail, the parts which are merely ornamental, and without necessary connexion with the architectural forms. Hence the yet unsettled questions respecting the Naos and its covering. Stuart inferred, from various circumstances, that the temple was hypæthral, "that is, with two interior ranges of columns dividing the cella into three aisles; of these, the two next the walls alone were roofed, and that in the centre exposed to the heavens." On the contrary, Mr. Wilkins, whose words we have adopted in the brief explanation just given, contends that there are no adequate grounds for Stuart's inference. Colonel Leake again, apparently supported by the investigations of Mr. Cockerell, inclines to the former opinion, which is also supported by the recent editor of "The Antiquities of Athens." The plates in the present work

and comprising sculptural decoration alone for one edifice, exceeding in quantity that of all our recent national monuments; consisting of a range of eleven hundred feet of sculpture, and containing on calculation upwards of six hundred figures, a portion of which were colossal, enriched by painting and probably golden ornaments. Here has been really verified the prediction of Pericles that, when the edifices of rival states would be mouldering in oblivion, the splendour of his city would be still paramount and triumphant."

are in accordance with the hypothesis which classes the Parthenon among hypæthral temples.*

The decorations of this sumptuous edifice were of the richest and most perfect design and execution. Both pediments were charged with sculpture of unrivalled excellence: the metopes of the exterior entablature exhibited a succession of ninety-two groupes in high relief: and the frieze, which surrounded the cella and vestibules, was adorned in its entire length of more than five hundred feet, by a representation in low relief of the Panathenaic procession. But the great ornament of the temple was the chryselephantine statue of the goddess, which stood in the cella: framed of the most costly materials, and wrought by the very hand of Phidias, this wondrous work had but one rival, and that was by the same master, and of the same materials.

This structure is sometimes called the Hecatompedon, either from its actual dimensions, or from the

* Mr. Wilkins, in his latest publication, the *Prolusiones Architectonicæ*, has done us the honour of commenting on this paragraph. After speaking in complimentary terms of our "Epitome," he goes on to reassert his opinion, though without repeating or reinforcing his previous reasoning on the subject. His intention seems to have been, by showing that Colonel Leake and Mr. Cockerell had, *in other cases*, given up positions maintained by them conjointly, to intimate that they were, therefore, likely to be wrong *in this*. We cannot think that the inference is quite legitimate, but as we have not, by any means, a tenacious feeling in the matter, we are content to leave it without controversy.

distinctive name of an ancient temple which formerly occupied the same site.

PLATE XXIII.

Fig. 1. Plan of the Parthenon.—In the western front, marked A, A, was the principal entrance. B is the pronaos; and it may be well, in this place, to point out a peculiarity in its arrangement. In the common construction it was usual to prolong the pteromata, or lateral walls, until the antæ were on the same alignment as the intervening columns: in this instance, however, the colonnade is complete, covering the antæ by the exterior column on either flank; the general proportions being preserved by reducing the pteroma from its usual extent, to something of little larger dimensions than a buttress. C, the cell, cella, or naos, where stood the statue of Pallas. As this plan is strictly a reduction of that of Stuart, it is necessary to remark that the columns indicated as supporting the central portion of the cella are of uncertain origin. It will be observed that they are of a diameter much less than that of the colonnade of the pronaos, although this has a slenderer shaft than that which belongs to the pillars of the portico. Stuart supposed that the interior parallelogram was composed of a double range, the lower supporting an entablature, as a sort of stylobate to an upper and shorter tier, on which rested the roof of the aisles, the centre being hypæthral. It does not, however, appear that the columns thus indicated belonged to the original building; Fauvel believed that they were of the Lower Empire, and the entire question may be considered as yet open. D, opisthodomus, represented in Stuart's plan as having been supported by six columns: a more recent and minute examination by Mr. Cockerell gives but four. It will not appear surprising that all these uncertainties should present themselves, when it is recollected that a new mosque was built by the Turks within the walls and with the very materials of the temple itself; and that further dilapidations have been committed to a great extent. The ravages of war have combined their devastation with these larcenies of peace: in 1687, when the Acropolis was besieged by the Venetians under Morosini, a shell fired a powder-magazine which occupied the interior of the temple, and the principal mischief done to the ornamental parts seems to have been the effect of this explosion.

Fig. 2. Transverse section of the portico, of which the columns are removed, for the purpose of showing those of the pronaos, which stand on a platform raised two steps above that of the portico.—They support an architrave, surmounted by the Panathenaic frieze, which is continued round the temple.

PLATE XXIV.

Elevation of the Parthenon, with the sculpture of the frieze and pediment restored.

The architrave is ornamented with shields, of which one is suspended over each column; in the intermediate spaces are inscriptions. It should be observed that this arrangement is, in a great degree, arbitrary.

PLATE XXV.

Side view, combining the advantages of an elevation and a section.—The wall is broken away from the central portion, so as to exhibit the opisthodomus, and the hypæthral cella, with its double range of columns and the Phidian statue of Minerva Parthenos.

PLATE XXVI.

Fig. 1. Capital and entablature of the columns of the portico.

Fig. 2. Mouldings of the capital, on a larger scale.

PLATE XXVII.

Fig. 1. Capital and entablature of the pronaos and posticum.

Fig. 2. Capital of the antæ of the posticum, and section of the entablature.

Fig. 3. Mouldings of the capital of the antæ, on a larger scale.

PLATE XXVIII.

Pediments of the eastern and western fronts, restored.

The workmanship of this noble edifice exhibits the exquisite finish which distinguishes the best period of Greek art. No cement was used in the construction, but the masonry is fitted with the utmost accuracy, and held together by iron cramps run with lead. The cylindric blocks which form the columns, have their upper and lower surfaces adjusted and secured by wooden pins and plugs. Certain vacuities and apparent neglıgences in different parts of the building, though they may be accounted for without violence to sound principles, suggested to the artist Lusieri, the singular and somewhat whimsical notion that the Greek masons were knaves, and that Pericles had been cheated by his workmen.

The subject of colour in its application to architectural effect, has already been slightly touched, and it would involve too great an extent both of detail and discussion to follow it out in this place. It has, however, evidently been so much employed by the Greek architects as a legitimate resource, that it would be inexpedient to pass it by altogether. The marks of paint are still clearly visible on many of the ornamental parts of the Parthenon. The capitals of the antæ; the members of the architrave and frieze; the mouldings of the pediments; were severally adorned with the designs usually distinguished as the Fret—the Palmette—the Egg-and-dart. One portion of the frieze was marked with zig-zag stripes; and the lacunaria were doubtless enriched with colours and gilding. The sculpture was probably, perhaps advantageously, relieved by a light-blue ground, and the figures, draped or nude, might possibly be distinguished by appropriate tints.

THE ERECHTHEUM.

PLATES XXIX. XXX. XXXI. XXXII. XXXII. *bis*, XXXIII. XXXIV. XXXV. XXXVI. XXXVII. XXXVIII. XXXIX.

NORTH of the Parthenon, at the distance of about one hundred and fifty feet, are the remains of three contiguous temples. That towards the east was called the Erechtheum; to the westward of this, but under the same roof, was the Temple of Minerva, with the title Polias, as protectress of the city; adjoining to which, on the south side, is the Pandrosium, so named because it was dedicated to the nymph Pandrosus, one of the daughters of Cecrops.—*Stuart*.

The passage in Pausanias on which Stuart founded his opinion that this beautiful but irregular structure had a three-fold dedication, scarcely sanctions the conjecture; in fact, the rambling and discursive manner of the Greek traveller requires, here as elsewhere, the aid of exact local investigation, before it can be made to support anything in the shape of definite result. Subsequent examination has brought to light important circumstances, which were not accessible in the time of Stuart. Mr. Inwood devoted much skilful labour to a personal inspection, and Mr. Wilkins has made proof of admirable scholarship in his examination of one of the most singular and instructive among the remaining inscriptions of antiquity. It is not the least of the many peculiarities of this temple, that it never received the last finish, and the inscription in question contains the particulars of a minute professional survey of the unfinished parts, conducted under the direction of a regular architect (Philocles), employed by the local authorities.

From all these sources of information it appears, in the judgment of the ablest palæologists, that the entire building was a double temple, of which the eastern division was consecrated to the worship of Minerva; and the western, including the northern and southern porticoes, was sacred to the deified daughter of Cecrops. On the same site had formerly stood the Temple of Erechtheus; and from this circumstance, as well as from the fact that his altar still remained, the entire building retained the name of the Erechtheum. Within the sacred enclosure were preserved the holiest objects of Athenian veneration, among which the most precious were the olive of Minerva and the fountain of Neptune, both which sprung up at the bidding of those divinities, when there was contention among the Gods, concerning the guardianship of Athens.

Here, too, was the oldest and most deeply venerated of the three famous statues of the Athenian Goddess. The "Great Minerva,"—the Minerva Promachus, so called from its martial bearing—was of bronze, the work of Phidias. This colossal figure was in part visible to the navigator who neared the port of Athens; and it was the first object that met the eye on entering the Acropolis. The second of these celebrated statues was the Minerva of the Parthenon, also by Phidias, wrought in ivory and gold, the noblest example of the *toreutic* art. But the admiration awakened by these sublime productions of human genius and skill, was a mixed and imperfect feeling, compared with the religious awe which impressed the worshipper who bowed before the Minerva of the Erechtheum, a figure carved in olive-wood, probably of inferior workmanship, but of which the legend affirmed that it fell from heaven. "This," in the eloquent language of Mr. Wordsworth, "was the Minerva Polias: the original Minerva of Athens; the Minerva who had contested the soil of Attica with Neptune, and had triumphed in the contest: the Minerva of the Acropolis, and of the temple now before us. Inferior to the other two in value of material and beauty of execution, she was regarded with greater reverence. Hers was emphatically the *ancient* statue: to the Minerva *Polias* it was, and not to the Minerva of the *Parthenon*, that the Panathenaic peplus—the embroidered fasti of Athenian glory—was periodically dedicated."

It may serve to simplify the somewhat complicated form of this temple, if the porticoes on the flanks be for the moment discarded. In that case, it is well observed by Mr. Wilkins that "the plan would be simply that of the kind of temple termed by Vitruvius prostyle; that is, with a portico in the principal front only, and no peristyle. If to a temple of this description two porticoes be added at the western extremities of the flanks, a general idea of the plan of the building may be formed." It will, however, be better to leave all further ichnographical detail to the explanation of the plates, where the description and the exemplification may stand side by side.

Nothing can go beyond the workmanship of this temple. The ornaments, throughout, are of the most finished execution, and the sculptors seem to have derived all possible advantage that was afforded them by a material which admitted of being wrought with "the delicacy of an ivory cabinet." "In this beautiful specimen of the Ionic order," observes Colonel Leake, "the Athenians seem to have been ambitious of excelling their Asiatic brethren in their own peculiar order of architecture, by the addition of new and elaborate ornaments, imagined with the utmost ingenuity and elegance of taste, and executed with a sharpness and perfection, which it could hardly have been supposed that marble was capable of receiving." The sculptured necking of the columns is said by Mr. Wilkins to have been "observed in no other known instance of the Ionic order: the volutes are beautiful in design, and most exquisitely wrought." It has been said of the volutes, that they were not struck from centres. This temple supplies an additional example of the rule which obtained among the architects of Greece, that there should be no similarity between the capital of the column and that of the antæ; contrary to the practice of the Roman builders, whose system it was to harmonise the respective features as far as possible. In the present instance the antæ present no trace of the volute. It deserves notice, too, as a peculiarity in the construction of this edifice, that the frieze and part of the pediment is faced with thin slabs of a grey limestone, which is, in the inscription already referred to, called "Eleusinian stone."*

* Mr. Woods, in the "Letters of an Architect," describes this "Eleusinian stone" as a *black* marble, and its present grey tint as the effect of weathering. He suggests that it was probably enriched by ornaments of gilt bronze.

HUS far the former publication, and this abstract, brief as it is, was the result of somewhat more research than would be anticipated by readers inexperienced in the difficulties which beset the architectural student in his efforts to combine and interpret the "rich relics" of antiquity. He has to correct the careless and inadequate draughts of the mere describer; to search out the casual references of the historian and the poet; to repair the errors with which time and negligent transcription have marred the record; and to extract from these imperfect and sometimes conflicting elements, a clear and harmonising explanation of mutilated and scattered fragments. A task, this, never easy, sometimes impracticable, and, if not positively hopeless in the present instance, requiring, it should seem, yet further investigation before it can be considered as satisfactorily completed. A careful review of the facts and authorities has appeared to justify a repetition of the statements and citations as given in the first edition. With the view, however, of rendering the details more complete, the passages in Pausanias which relate to this splendid but singular structure, may be advantageously laid before the reader, in free but fair translation; and in addition to this,

it is due to the learned and laborious investigations of Mr. Wilkins, that their latest results should be recorded here.

"Before the vestibule of the Erechtheum, stands an altar to Jupiter Hypatus (Highest), on which they sacrifice nothing that has life, but placing cakes, they do not sanction even the use of wine. In the very entrance are altars; one to Neptune, on which also offerings are, by command of an oracle, made to Erechtheus: another is consecrated to the hero Butes: the third to Vulcan. On the walls are paintings referring to the descendants of Butes. The edifice is double, and there is in it sea-water in a well (or reservoir) . . . . when the south wind blows, a sound is heard as of waves. On the rock is impressed the form of a trident. These things, it is said, are testimonials of Neptune's contest concerning Attica *with Minerva.*

"Sacred to Minerva, indeed, are the city and the entire region; for even in those communities where other deities are held in special honour, none the less is Minerva reserved."

Pausanias then goes on to speak of the wooden image to which we have already referred as of peculiar sanctity, and which was consecrated and placed in the Acropolis by the common consent of the Attic Demi; an act that appears to have been the solemn and religious recognition of their national union under the supremacy of Athens. He next mentions the

Golden Lamp, the master-work of Callimachus, burning night and day, yet requiring to be replenished with oil only once in the year, the unconsumable wick made of Carpasian flax, and the brazen Palm-tree rising to the roof and carrying off the smoke of the Lamp. There, too, was the ancient Hermes, the gift of Cecrops, hidden in myrtle-leaves,—a folding seat,* the work of Dædalus—the cuirass of Masistius, and the scymitar of Mardonius. Concerning the identity of these barbarian trophies, however, Pausanias intimates some doubt. The armour of Masistius, he thinks, may be taken as genuine, since that gallant officer was cut down in the act of charging the Athenian horsemen; but he is puzzled to know how the sabre of the commander-in-chief, who was slain in fight with the Lacedæmonian division, could find its way to the Acropolis of Athens. The difficulty does not seem formidable: though the Spartans were victorious in the field, they failed before the fortified camp of the Persians; and the Athenians, who carried it gallantly by assault, would probably find the weapon in the richly-furnished tent of Mardonius.

In this temple, moreover, is preserved the sacred Olive, the memorial of Minerva's victory over Neptune when contending for Attica. Concerning it they hold the tradition that when the Persians fired the city, this tree was consumed; but that in the course of the same day, it threw out a fresh shoot to the height of

* Perhaps a kind of chariot.

two cubits. Near the temple stands a statue of an aged female, a priestess of Minerva, well executed: and at a short distance are two large bronze figures of men in combat: one of Erechtheus, the other of Immaradus, who was slain in the fight. There are in the citadel several ancient statues of Minerva, still unmutilated, but so black and burnt as to be incapable of resisting the slightest violence. In their present state they are most impressive memorials of the conflagration in which all that was in Athens perished when the Persians took possession of the city.

Mr. Wilkins, to whom this work has been already much indebted, and whose close and critical observation is quite otherwise instructive than the vague and discursive manner of Pausanias, made this temple the object of long and successful examination. His commentary on the curious inscription to which reference has been already made, gave a new direction to the study of architectural antiquity; and a dextrous application of conjectural criticism gave, in one important instance at least, consistency to the shattered text of Vitruvius. In 1837 he published his "Prolusiones Architectonicæ:" here he resumed his favourite subject, and the larger portion of a thin quarto is taken up by elucidations of the Erechtheum and its inscribed marble. With this last we have nothing to do, since it is quite foreign from our subject to discuss the technicalities of art; but the earlier part, as expressing the ultimate opinion of an eminent man, claims from us a brief notice. After an ingenious manipulation of a stubborn passage in Xenophon, followed by a brief description of the Acropolis, he goes on to detail the ichnography of the building. With Stuart and others, he divides the edifice into three parts, of which that which was entered from the eastern or hexastyle portico was the cella of the temple of Minerva Polias. This was separated by a massive wall from the Pandroseum and its pronaos, to which the tetrastyle portico on the flank, or rather shoulder, of the structure, afforded the only access. To the opposite *humerus* was attached the famous Caryatid "prostasis," constructed, as generally supposed, for the purpose of affording light and air to the sacred Olive which grew within it. This theory is disputed by Mr. Wilkins, who supposes the Neptunian spring to have occupied this part of the building, and the tree of Pallas to have flourished in perpetual verdure within the pronaos itself, access of light and circulation of air being obtained from the windows in the intervals of the engaged columns in the western front.

PLATE XXIX.

Perspective view of the Erechtheum, reduced from the large and elaborate restoration by Mr. Inwood.* In this draught there is much detail for which no

* We are indebted to the liberal courtesy of this gentleman for permission to copy his plate.

direct authority can be given, though there is probably none for which plausible reasons might not be assigned.

PLATE XXX.

Fig. 1. Plan.—A, "Temple of Erechtheus, or of Neptune, in which was the well of salt water, and the altars of Neptune, of Vulcan, and of the hero Butes; before it stood the altar of Jupiter the Supreme. B, The temple of Minerva Polias, perhaps the Cecropium of the Dilettanti inscription. C, The temple of Pandrosus, in which was the olive produced by Minerva, and the altar of Jupiter Herceus. D, The portico, common to the temple of Minerva, and to that of Pandrosus."

As already intimated, this arrangement, which is given in Stuart's own words, has been considerably modified by the results of subsequent research. It appears nearly certain that the eastern division, marked A, is to be taken as the cella of the temple of Minerva Polias, while the remainder, including the tetrastyle portico and that of the Caryatides, was known as the Pandroseum. The Cecropium is, with the highest probability, supposed to have been a building near, but separated from, the stylagalmatic entrance.

Among the singularities of this structure is to be noted the difference of level, both within and without. The chamber, marked A, has for its floor a platform higher by ten feet than that of the other division; and, if the suggestion of Mr. Wilkins be correct, this arrangement was made for the purpose of constructing a sepulchral chamber beneath, containing, probably, the tomb of Erechtheus.

Fig. 2. Elevation of the tetrastyle, or northern portico.

It is remarkable that in each of the three ranges of Ionic columns which are connected with this temple, the intercolumniations differ. The engaged columns of the western front are nearly eustyle, standing apart something less than two diameters and a fourth; this, according to Vitruvius, is, as the term imports, the most perfect of all the columnar systems for beauty and strength. The columns of the eastern or hexastyle portico may be considered as systyle, having an interval of little more than two diameters. The northern portico is diastyle, with an intercolumniation of nearly three diameters.

The wall separating the western chamber from the tetrastyle and caryatid porticoes, was ascertained by Mr. Inwood to be of later construction than the rest of the building.

PLATE XXXI.

Entablature, capital, and base, of the tetrastyle portico. These columns are in several respects more highly ornamented than those of the other fronts.

PLATE XXXII.

Doorway to the tetrastyle portico.—The details of this magnificent entrance appear to have been chiefly taken from Inwood (*Erechtheion*). It still exists in a state "nearly perfect," though in Stuart's time it was not accessible; the portico having been walled up by the Turks, and made to serve the purpose of a powder-magazine. Mr. Wilkins was not more fortunate than Stuart, but Mr. Inwood's visit was under better auspices. Mr. Donaldson, in his elaborate and useful work on "Ancient Doorways," has bestowed great and apparently successful pains on this "beautiful gate." His draught does not exhibit the rich crowning ornament of the Hyperthyrum.

PLATE XXXII. (*bis.*)

Front, side-view, and section of the consoles supporting the hyperthyrum of the preceding doorway.

PLATE XXXIII.

Elevation of the Erechtheum, showing the eastern or hexastyle portico, with its pediment; the side elevation of the northern or tetrastyle portico; and the stylagalmatic * portico of the Caryatides. By the mismanagement of the engraver, the relative positions of these porticoes are reversed.

* This rather formidable compound was first applied to the Figure-column by Mr. Wilkins, and was adopted in this work as having at least the merit of being accurately descriptive. The German critics, however, object to it as "barbarous." Mr. Wilkins sarcastically alludes to this as a "kind of Chinese refinement" on the part of Teutonic philologers, but expresses his perfect willingness to accept, "for the sake of euphony," the term Coræatic, from Κοραι —virgins or young females—a word specifically applied, in the Erechthean Inscription, to the statue-pillars (Bild-saülen) of the Caryatid Prostasis.

PLATE XXXIV.

Fig. 1. Base, capital, and entablature of the eastern portico.

Fig. 2. Capital of the antæ to the western front.

Fig. 3. Base of the same.

The reader for scientific or practical purposes is referred to the plates of Mr. Inwood's great work. They are boldly sketched, drawn to scale, and of large proportions. Many particulars, too, of considerable value, are inserted. We do not recollect whether he was the first to observe the insertion of coloured stones or glass in the platband of the base, but they do not appear to have been noticed by Stuart, and we are not aware of any previous authority for the fact.

PLATE XXXV.

Plan of the above, in reverse, with sections of the capital, and contour of the volute.

PLATE XXXVI.

Capital and base of the columns of the western front: outline of the volute.

PLATE XXXVII.

Fig. 1. Section, showing the interior of the western wall, connecting the tetrastyle and caryatid porticoes. In the windows are indications which lead to the supposition that they were closed with somewhat of transparent material.

Fig. 2. Section of the wall of the western front.

Fig. 3. Elevation of the caryatid portico.

PLATE XXXVIII.

Details of the entablature, capital, and stylobate, of the portico of the Caryatides, with portions of the figure.

PLATE XXXIX.

Fig. 1. Capital of the antæ of the same portico, with section of the ceiling or soffit.

Fig. 2. Plan of the soffit.

ODEUM OF REGILLA,

OR,

THEATRE OF HERODES ATTICUS.

PLATE XL.

STUART, Wheler, Pococke, and more recently Mr. Haygarth, have assigned these ruins to the Theatre of Bacchus. Nothing, however, can be more complete than the chain of evidence and argument, by which Leake, Wilkins, and Kinnaird, have established their claim to be considered as the remains of the Odeum, or Music Theatre, erected by the wealthy and munificent Herodes Atticus, in memory of his wife Regilla.* It is probable that the removal of the rubbish which has fallen from

* This generous patron of the Arts appears to have been one of the fairest characters of ancient times. So far as public records may be trusted in attestation of private and patriotic excellence, his life was of unimpeached integrity. As a friend and citizen, his kind and

the higher ranges of the circle, and accumulated in front of the proscenium, would bring to light many an interesting and illustrative fragment; but Stuart was hindered in his examination by the jealousy of the Turks, and the researches of succeeding visitants have not penetrated below the surface.

This theatre, which is said by Pausanias to have excelled all similar structures in Greece, was hollowed out of the rock on which stood the Acropolis; the seats were also part of the solid mass, but the whole was cased with marble, as were also the walls and ornamental portions of the proscenium. Hence it was not practicable to form those numerous corridors and vomitoria which gave such free access to all parts of a Roman theatre. "There appears," says Mr. Wilkins, "to have been only two ranges of seats; the præcinction, or passage separating them, may be still distinguished. The only approaches to the theatre were at the horns of the auditory, where the staircases communicating with the præcinctions are remaining."

The general dimensions are given on the same authority, as follows: "The cavea is the greater

liberal disposition has the testimony of an inscribed marble which remains to our own day; that his domestic affections were strong and lasting, may be inferred from the splendid memorial of conjugal tenderness which he consecrated to the name of his deceased wife. The inscription is in the usual terse and expressive style of classic eulogy, "To the High Priest of the Cæsars—Tiberius Claudius Herodes, the Marathonian—on account of his goodwill and beneficence toward his country."

segment of a circle, whose radius is one hundred and twenty-four feet . . . . . the front of the scene recedes twenty-five feet from the chord line . . . . . the extent of the scene, exclusive of the two returns, is one hundred and seventeen feet."

A reference to the diagram (Plate xl.) will, however, make the forms and arrangement much clearer than pages of vague description, and the plan will be found to show with perfect distinctness, both the particulars of the Odeum of Regilla, and the general disposition of a Greek theatre; the proscenium, with its connected apartments; the orchestra, immediately in its front; and the cavea, or coilon, with its præcinction and staircase.

The principles which regulated the construction of the Greek Theatres, have obtained much and skilful elucidation from the antiquarian and the architect; and it is a matter of no small regret that the intention of furnishing the reader of this volume with a general view of the subject, has been unavoidably laid aside. The truth is, however, that mere summary has been found so vague, partial explanation so unsatisfactory, and the necessity for multiplying diagrams so inevitable, that the design has been abandoned. The student, whether amateur or professional, is referred to the Vitruvius of Mr. Wilkins for an able digest of whatever has been ascertained or suggested on this important section of architectural science, illustrated by plans and sections.

Mr. W.'s restoration of the Theatre of Herodes Atticus differs considerably from the more partial exhibition of Stuart; and the insertion here of his analysis will add materially to the value and interest of this sketch. "There are no traces," he states, "of staircases between the cunei to be discovered in the ruins of this theatre; but in the wall which surrounds the upper precinction there are remains of niches, or recesses, which, like the real doorways in other theatres, were probably opposite to the ascents. Upon this supposition, their disposition would correspond very nearly with that which we are desired by Vitruvius to adopt; for the ascents would begin from the angles of the squares inscribed in the circle of the orchestra... The cunei on the right and left of the scene were of greater extent than the others; a similar inequality is likewise apparent in the plan of the theatre at Tauromenium; and seems to have been dictated by the propriety of giving the same facility of access to all the cunei. The staircases at the extremities of the cavea afford access to the seats of the two cunei only contiguous to them; whereas each of the others communicates with those of the two cunei which it separates: so that were the cunei of equal extent, the facility of approaching the seats of those next the extremities would be greater than what was afforded to the others."

"The theatre having been excavated in the side of the rock of the Acropolis, there were no other approaches to the precinctions than those at the back of the scene."

"The orchestra is the segment of a circle greater than a semicircle. It is probable that Vitruvius had a theatre similar to this in view when he represented the orchestra of the Greek theatres as formed by arcs described from three several centres. The form of the orchestra in the theatres at Stratonicea, Miletus, Laodicea, and Iassus, was a considerable portion of a circle."

THE CHORAGIC MONUMENT OF THRASYLLUS.

PLATES XLI. XLII.

A SINGULAR building, which dates a century lower than the Periclean age, and exhibits much that is impressive and well adapted to the purposes of the architect and sculptor. The lower portion of the edifice has evidently been considered as strictly subordinate to the entablature and attic; and these have been disposed with specific regard to the great feature of the entire arrangement, the noble statue which occupied the centre of the crowning platform. This statue, now in the British Museum, has lost the head and arms; but enough remains to leave it little doubtful that it represented Bacchus in some of his various aspects. Stuart supposed it to represent a female, Dr. Chandler guessed that it might be a Niobe, and Mr. Wilkins describes it as "a colossal figure of the female Bacchus in a sitting posture." Dr. Clarke imagined that he could trace sufficient indications to justify him in assigning the statue to the "bearded" or Indian Bacchus. Amid all these variations, the balance of evidence is decidedly in favour of the son of Semele, though in which of his manifold disguises may admit a question.

The whole structure forms an architectural front to a cave of small extent, now dedicated as a chapel to Our Lady of the Grotto, Panaghia Speliotissa.* This cavern stood immediately above the Dionysiac Theatre, and now serves as an important indication of the site of that celebrated scene of Choragic competition. The plan and restored elevation (Plate xli.) show distinctly the general aspect of the building, and the way in which it is connected with the recess. The design consists of an attic, broken in the centre by steps leading to the platform; the two wings, or flanking members, thus formed, present the appearance of a pedestal, or stylobate, with cornice, die, and plinth. This superstructure rests upon an entablature supported by two antæ and a central pillar or pier, quadrangular like the antæ, but differing from them in its proportions, and in the profile of its

* As an open cave was ill adapted to the ritual of the Virgin, the front has been walled up, and a view of the interior may be seen in the first volume of Dodwell's "Classical and Topographical Tour in Greece." The effect is picturesque enough, but the contrast between the mean furniture of the Chapel, and the noble character of the Monument, is singularly impressive.

mouldings. The first view of all this suggests the idea of irregularity, and makes it probable that there has been some alteration of the original design: this probability is strengthened by the fact that the attic is made up of a different marble from the pure Pentelic of the entablature and its supports. It might, too, be supposed from the unusual and rather awkward appearance of the middle prop that it had been an after-introduction, for the purpose of meeting the additional pressure of the statue: this inference, however, is disproved by the fact that the architrave is not of one continuous stone, but is jointed immediately over the pier.

Some light may be thrown on these difficulties by a reference to the history of the monument, as preserved in the three inscriptions still clearly legible on the architrave and attic. Of these, the most ancient is that on the centre of the architrave, recording the choragic victory of Thrasyllus of Deceleia, in the archonship of Neæchmus. The second shall be given entire, as a specimen of the style of these memorials, and as an illustration of the manner in which the performances themselves were got up:—

> The people gave the games: Pytharatus was Archon:
> Thrasycles, son of Thrasyllus, a Decelian, was Agonothetes:
> The Youths of the tribe of Hippothoon gained the victory:
> Theon, the Theban, played the flute:
> Pronomus, the Theban, composed the piece.

The other inscription records a contemporaneous

triumph achieved by the same individual. Mr. Wilkins, in his valuable but somewhat paradoxical "Atheniensia," has read these documents strangely. He affirms the double victory to have been gained in the earlier archonship of Neæchmus; whereas the inscriptions assign the first only — that of Thrasyllus — to his magistracy, and the two subsequent victories of Thrasycles to the much later archonship of Pytharatus. In addition to this error, Mr. W. gives 328 as the date of the presidency of Neæchmus, whereas the table of Olympiads, as published by Playfair, places it in 320, and this is followed by Leake and Gell. Pytharatus presided 271 B.C. Independently, however, of all discussion respecting dates, the mere fact that these inscriptions relate to more than one victorious Choragus, may assist in explaining some of the irregularities in the construction of the monument, and make it probable that the upper portion, with the statue, was an addition to the primary design.

The execution, although not to be compared with the exquisite workmanship of the monument of Lysicrates, is good, and the statue is "the work of an excellent sculptor."

The cave is described by Dodwell, in the work to which reference has been made in a foregoing page, as having been "originally formed by Nature," and subsequently "enlarged by Art." "It penetrates about thirty-four feet under the rock, and its general breadth is twenty feet. The only antiquities that it contains are a few blocks of marble, a small columnar pedestal, perhaps for a tripod, and a fluted columnar altar . . . . Here is also an Ionic capital of small proportions and coarse workmanship, with some appropriate paintings of the Virgin of the Cave. It receives a dim and mysterious light, through two small apertures in the modern wall, by which a singular and picturesque effect is produced."

PLATE XLI.

Plan and elevation of the Choragic monument of Thrasyllus.

PLATE XLII.

Fig. 1. Details of the entablature and capital.
Fig. 2. Vertical section of the entablature.
Fig. 3. Profile of the attic.

THE PROPYLÆA.

PLATES XLIII. XLIV. XLV. XLVI. XLVII.

OF all the examples that yet exist of Athenian magnificence this splendid structure may be considered as the most thoroughly characteristic. It was, when in its perfect state, the admiration of Greece; and Epaminondas—whether in menace or in metaphor may be doubtful—proposed its forcible removal to the Cadmean citadel of Thebes. "This work," says Colonel Leake, "the greatest production of civil architecture in Athens, which equalled the Parthenon in felicity of execution, and surpassed it in boldness and originality of design, was begun in the archonship of Euthymes, in the year before Christ 437. . . . It was built under the directions of the architect Mnesicles, who completed it in a few years." Certain writers of antiquity have rated its cost at the enormous sum of two thousand and twelve talents, an amount hardly credible in itself, and quite at variance with the calculations of other and more trustworthy authorities.

Stuart was prevented from completing his survey of the Propylæa, by circumstances which compelled him to hasten his departure from Athens. Happily the deficiencies thus occasioned were to a considerable extent supplied by the subsequent visit of Revett, in whose drawings, admirably accurate in general, there are, however, a few errors, which will be corrected in the following details. Many particulars, which could not be ascertained from the existing remains, have been, since the publication of Stuart's great work, obtained from a very singular and unexpected source. It seems that there was a nearly exact duplicate of the Athenian Propylæa. The entrance to the sacred inclosure of the Temple of Ceres, at Eleusis, was modelled, in after times, upon the structure of Mnesicles, and by marvellous good fortune, that which is wanting in the one has been found still extant in the other.

By referring to the plan in Plate xliii. it will be seen that the system of buildings which usually passes under the general designation of "The Propylæa," consists of a centre and two projecting wings, the fronts facing inward, and the whole forming three sides of a quadrangle. Of these three divisions the first only is the Propylæa, or ornamental approach to the five gates by which the Acropolis was entered: the colonnade on the right adorned the temple of the Wingless Victory: the opposite division contained, in an interior chamber, the celebrated paintings of Polygnotus. Concerning the entire structure a rather whimsical hypothesis has been started by Colonel Leake. He has argued, in his valuable work on the Topography of Athens, very ingeniously in favour of the supposition that this was altogether a military arrangement, an important post strongly fortified with a special view to the system of attack and defence prevalent among the Greeks. It were a mere waste of time and topography to deal argumentatively with suppositions so utterly untenable: and it may perhaps be equally idle to offer another suggestion, which shall, however, be hazarded, though without laying undue stress on either its nature or its probability. The western face of the Athenian citadel was its only accessible side; and of this face the Propylæa occupied the centre. There can be no question of the fact that, as on all weak points of a fortification, barricades were multiplied in advance of this post, nor that when these were either stormed or beaten down, the assailants stood before the last and weakest defence of the garrison, and had only to force the doors of a simple wall, in order to become masters of the interior. In effecting this, they would derive great advantage from the shelter afforded them by the strong stone-roof of the Propylæa, and find themselves protected from injury, just at the very moment when the last and most desperate efforts were to be made for their exclusion. Such a defensive system is in opposition to all sound principles of military construction, and it seems necessary to seek its explanation in some more plausible theory. Now is it an improbable conjecture in this case, that religious feeling may have had something to do with the matter, and that these temple-like structures might be intended as an appeal to the divine Protectors of Athens? Failing all other defences, there was yet a hope that the hitherto successful enemy would be arrested at the consecrated barrier.

Considerable uncertainty prevailed in the time of Stuart respecting the existence of a carriage-way through the Propylæa; and the two circumstances which seemed to prove the affirmative were, first, the width (ditriglyph) of the central intercolumniation; secondly, the introduction of chariots in the Panathenaic procession, as represented on the frieze of the Parthenon. Subsequent examination, under more favourable auspices, has confirmed this inference, by the discovery of evident traces of the skilfully-constructed inclined plane, along which carriages passed on to the interior of the Acropolis.

Colour and gilding appear to have been extensively employed in the decoration of this splendid structure.

Among the Elgin marbles in the British Museum may be seen a block still retaining traces of the interior painted cornice.

"Here," in the glowing language of Mr. Wordsworth, "above all places at Athens, the mind of the traveller enjoys an exquisite pleasure. It seems as if this portal had been spared in order that our imagination might send through it, as through a triumphal arch, all the glories of Athenian antiquity in visible parade . . . . . It was this particular point in the localities of Athens which was most admired by the Athenians themselves: nor is this surprising: let us conceive such a restitution of this fabric as its surviving fragments will suggest,—let us imagine it restored to its pristine beauty,—let it rise once more in the full dignity of its youthful nature,—let all its architectural decorations be fresh and perfect,—let their mouldings be again brilliant with their glowing tints of red and blue,—let the coffers of its soffits be again spangled with stars, and the marble antæ be fringed over as they were once with their delicate embroidery of ivy-leaf . . . . . and then let the bronze valves of these five gates of the Propylæa be suddenly flung open, and all the splendours of the interior of the Acropolis burst upon the view."

PLATE XLIII.

Fig. 1. Plan of the Propylæa.—A, the Propylæa, properly so termed. B, chamber, ornamented by the paintings of Polygnotus. C, temple of Wingless Victory. It will be seen by the inspection of this diagram, aided by a reference to the longitudinal section which is given, fig. 2, on the same plate, that the Propylæa (or rather, in strict application to the central structure, the Propylæum) consisted of a portico, a vestibule, and a posticum, or back portico, facing the platform of the Acropolis. These porticoes were Doric and hexastyle; the vestibule, or interior portion of the building, had its roof supported by six Ionic columns; but the pedestals, as represented in the plate, never existed: Revett's error was occasioned by misconception respecting the proportions of the shaft, and he had no opportunity of ascertaining the real circumstances by excavation. The marble beams which supported the ceiling and roof were from seventeen to twenty-two feet in length, and of proportionate solidity. The ceiling was richly carved and painted. Immediately behind the ranges of Ionic columns stood the terminal wall with five gates, diminishing in width and height as they receded from the centre opening, which was of sufficient dimensions to allow the passage of carriages; the relative proportions are indicated in the plan. The depth of the building, from the front to the wall, was forty-three feet, and beyond the wall was the posticum, adding eighteen feet to the above extent. The Temple of Victory, and the Painted Chamber, formed wings, nearly symmetrical externally, to the central Propylæum, although they were of different arrangement interiorly, as will be evident on reference to the plan. Each of them presented the exterior aspect of a pedimented portico with three Doric columns *in antis*. The entire of these constructions was of Pentelic marble.

Fig. 2 exhibits, in section, the longitudinal range of these buildings, with the different levels on which they stand. On the left are one anta and two columns of the right wing (B in the plan): next follow the Doric portico, and the Ionic vestibule, with the terminating wall; succeeding this, on the right, is the posticum, with pteroma, anta, and Doric column.

PLATE XLIV.

Elevation of the Propylæa.

PLATE XLV.

Fig. 1. Capital, architrave, and frieze of the central portico.

Fig. 2. Upper part of the shaft of the Ionic columns of the vestibule.

Fig. 3. Section of the external cornice.

PLATE XLVI.

View and section of the entablature and capital of one of the antæ of the Temple of Victory.

PLATE XLVII.

Fig. 1. Profile of the capital of the central portico.

Fig. 2. Section, on a larger scale, of the four annulets.

Fig. 3. Capital of one of the antæ.

This description of one of the most impressive productions of Athenian genius and liberality, cannot be more usefully terminated than by a paragraph of severe but judicious criticism from an article supplementary to Stuart, by Mr. Kinnaird:—

"Whether the design of the Propylæa, if viewed in concurrence with our modern opinions, founded on the experience of a greater variety of architectural composition, were conducted upon the principles of a correct taste, may be questionable, particularly as regards the juxtaposition of columns of different orders and altitude. It is, however, very evident that at the entrance to the sacred peribolus of the Acropolis, the ancient pictorial effect of this fabric, from its perspective combination as a foreground with the first distinct view of the surpassing Parthenon, must have excited equal admiration with the daring nagnificence of its construction. Of the force of this impression on the imagination, the full influence is made known to us by the recorded envy which threatened the removal of the edifice to the Cadmean citadel, during the Theban ascendancy, as well as by the existing proof of the imitation of its principal mass, both in form and dimensions, at the consecrated precinct of the mystic temple of Eleusinian Ceres."

CURSORY reference to the dimensions of the materials used in this structure will be found in a former page. As this, in all inquiries connected with architectural practice, is a matter of considerable interest, the following supplemental particulars, collected with characteristic diligence by Dodwell, though only in part relating to Greek construction, may be advantageously inserted here.

"The lintel over the middle gate is one of the largest masses of marble I have seen, being twenty-two feet and a half in length, four feet in thickness, and three feet three inches in breadth. It must accordingly weigh twenty-two tons. That of the second gate is sixteen feet ten inches in length, and three feet in thickness. That of the smaller gate is nine feet and a half in length, and three feet in thickness. The largest masses which remain in Greece are the beams of the Propylæan portico, the architraves of the Parthenon, the beams of the Erechtheion, of the Olympeion, and a block at the Pnyx at Athens, the lintels of the treasuies of Atreus at Mycenæ, and of Minyas at Orchomenos, and some stones in the walls of Tiryns and Messene.

"Some blocks of white marble are found in Italy, which vie with those of Greece; particularly two in the Colonna garden at Rome, which are supposed to have belonged to the Temple of the Sun. The largest is sixteen feet three inches in length, and nine feet and a half in thickness. Nor must I omit the architraves of the Pantheon, and of the temple of Antoine. The granite columns of the baths of Dioclesian, and of the forum of Trajan, as well as the Egyptian obelisks at Rome, are also examples of these stupendous masses. The architraves of the Temple at Selinus in Sicily are twenty-two feet in length. Tavernier mentions some blocks of an amazing size in a Pagoda at Golconda or Bagnagar.

"Chardin asserts that most of the stones of one of the temples at Persepolis are between thirty and fifty feet in length, and from four to six in height; and some of them are stated to be fifty-two feet in length.

"The columns of the famous temple at Cyzicum in Mysia, of one piece, were fifty cubits in height; but the largest mass that was ever moved by human means was the monolithal temple of Latona, at Butos in Egypt, which was a solid cube of sixty feet! There was another monolithe at Sais, of thirty-one feet and a half in length, twenty-one in breadth, and twelve in height. Wood informs us that in a wall at Balbec three continuous stones measure one hundred and ninety feet in length; the longest being sixty-four feet.

"The architectural remains of Egypt, which supply numerous examples of this colossal style, are too well known to require particular enumeration. . . . Among these gigantic masses our own Stonehenge must not be forgotten."

THE TEMPLE OF THESEUS.

PLATES XLVIII. XLIX. L. LI. LII. LIII.

HIS edifice, the least dilapidated among the remaining structures of ancient Greece, has in an especial manner excited the admiration of the most accomplished travellers.

Dodwell says of it that "this elegant building probably furnished the model of the Parthenon, which resembles it in the most essential points, though it is of nearly double the size. Indeed, the Theseion impresses the beholder more by its symmetry than its magnitude."

Sir William Gell describes it as "perhaps the most beautiful and best preserved monument of antiquity; and producing, notwithstanding its small dimensions of one hundred and four feet by forty-five feet, an inconceivable effect of majesty and grandeur."

Mr. Wordsworth is enthusiastic in eulogy. "Such," he exclaims, "is the integrity of its structure, and the distinctness of its details, that it requires no description beyond that which a few glances might supply. Its beauty defies all: its solid yet graceful form is, indeed, admirable; and the loveliness of its colouring is such, that, from the rich mellow hue which the marble has now assumed, it looks as if it had been quarried, not from the bed of a rocky mountain, but from the golden light of an Athenian sunset."

Stuart speaks of it more soberly, but not less impressively, as "one of the noblest remains of ancient magnificence, and at present the most entire."

This Doric temple is peripteral, with six columns in front, and thirteen on either flank. The portico fronts the east. The pronaos and posticum are formed by the continuation of the side walls, with two columns between the antæ. Every part of the building is of marble, and it rests upon a foundation of massive limestone blocks. All evidence concurs to establish the claim of this admirable structure to be considered as the temple of Theseus, erected in honour of that hero, and in expiation of the characteristic ingratitude with which the Athenians had repaid his eminent services and heroic deeds. Centuries rolled away before their hereditary guilt was thus atoned for; and the misdeeds of the olden time might yet have passed unrecognised, but for the miraculous appearance of a mighty vision, —the armed spectre of Theseus, scattering on the plain of Marathon the enemies of his ungrateful country. The Delphic oracle directed the recovery of his remains and their honourable burial. This pious duty was performed by the son of Miltiades, who found the hero's bones and armour in the island of Scyros, and brought them back to Athens, where they were welcomed with sacrifices and festivals: games were instituted and temples built, in honour of him dead, who, living, had been persecuted and driven forth to die by violence in a foreign land. His temple was not less revered than those of Pallas and Demeter: as a sanctuary it protected those who fled from the pursuit of law; and its peribolus was large enough to hold the military assemblies. Its erection dates thirty years earlier than the Parthenon.

Unlike the lavish decoration of the temple of Minerva, the Theseium was ornamented with a sparing hand, though the arrangement of the sculpture was so judiciously managed as to produce the greatest possible effect. Only the eastern or principal front of the temple appears to have been charged with figures: the posticum, indeed, had on the inner frieze a lively representation of the Feast of the Lapithæ, in which Theseus bore a distinguished part; but the entablature and pediment of the western portico exhibit no indications of similar adornment. The grand front, on the contrary, was filled with admirable sculpture; that of the tympanum has disappeared, but the frieze of the pronaos is covered with groupes, part of which are engaged in fierce conflict, while others seem to represent deities. The eastern portico exhibits the labours of Hercules on the ten metopes of the front, while four others immediately consecutive on either flank, apparently refer to Theseus, whose friendly alliance with the son of Alcmena is thus commemorated. It is by no means unlikely that these subjects were designed, if not executed, by the celebrated Micon. The remaining metopes of the sides have never been adorned with sculpture.

As the singular beauty of this temple is, in a great degree, the expression of its exact proportions, a deficiency in the foregoing illustrations may be advantageously supplied by the insertion of more ample details respecting the measurement of the principal parts. For this purpose a surer authority than that of Colonel Leake can hardly be found; and from him we learn that "the depth of the pronaos is greater than that of the posticum, and the depth of the portico of the pronaos is greater than that of the portico at the back of the temple; the two former measure together thirty-three feet, the two latter twenty-seven feet. The side porticoes of the temple are only six feet in breadth. The thirty-four columns of the peristyle, as well as the four in the two vestibules, are near three feet four inches in diameter at the base, and near nineteen feet high, with an intercolumniation of five feet four inches, except at the angles, where, as usual in the Doric order, the interval is made smaller, in order to bring the triglyphs to the angle, and at the same time not to offend the eye by the inequality of the metopes. The stylobate is formed of only two steps. The height of the temple, from the bottom of the stylobate to the summit of the pediment, is thirty-three feet and a half."

PLATE XLVIII.

Fig. 1. Plan of the temple. It presents an exact illustration of the Vitruvian arrangement of a peripteral temple. A, A, the two fronts. B, the pronaos. C, the cella, or nave. D, the posticum. E, E, the pteromata, or wings.

Fig. 2. Transverse section of the eastern portico, the front columns being removed to exhibit the columns, antæ, and lacunaria, of the pronaos, with the bas-relief which adorns the frieze.

PLATE XLIX.

Elevation of the eastern front.

PLATE L.

Fig. 1. Longitudinal section of the eastern portico and pronaos.

Fig. 2. Half the flank and half the longitudinal section of the temple, exhibiting the masonry, and the arrangement of the lacunaria.

PLATE LI.

Fig. 1. Section of the entablature over the columns of the portico, with details of the mouldings and lacunaria.

Fig. 2. Plan of the soffit of the architrave, and of the lacunaria.

PLATE LII.

Entablature and capital of the columns of the façade.

PLATE LIII.

Fig. 1. Details of the capital, on a larger scale.
Fig. 2. Plan of the flutings.
Fig. 3. Details of the four annulets of the capital.

The figures at the lower part of the plate are representations of ornaments painted in the soffit of the lacunaria.

This system, of which so much has been recently written, but to which it is yet hardly possible to reconcile our habits of taste and feeling, seems to have been applied extensively in the decorations of this temple. Not only have simple ornaments been painted in, but cornices, capitals, ceilings, and back-grounds, were thus distinguished. The carved metopes and friezes were, to all appearance, almost as completely subjected to the processes of colouring, as if they had been regularly elaborated by the painter on a plain surface: nothing seems to have been trusted to the *relief*, but the effects of light and shadow. The armour was touched with bronze and gold; the draperies were of various hues, among which blue, green, and red, may yet be distinguished; the sky still bears the traces of its proper tint, and everything appears to indicate that the Greeks either did not know—or, knowing, deliberately rejected—what would now be considered as an essential distinction between painting and sculpture. Without, however, attempting to decide, or even to discuss, this vexed and vexatious question, we may be assured that all this was executed, not by mere workmen but by artists; that it was carefully studied and judiciously applied; and it would, at least, follow that very striking effects must have been produced by the varying shadows of the day, and by the accidents of the atmosphere: the play of light and shade shifting from hour to hour would almost suggest the idea of life and motion; giving a character of reality to the groupes, not attainable by any other process. It will, however, still remain to determine whether this be art or artifice; and whether, in our estimate of these matters, we do not often suffer conventionalities and preoccupation to interfere with our reasonings concerning the legitimate objects and limits of Art.

A reference to Egyptian example, while it leaves the principle untouched, tends but little to relieve the historical difficulties, or even to illustrate the mere facts of the inquiry. If it be granted that the elements of Greek construction were derived immediately from Egypt—an hypothesis liable to many objections both circumstantial and theoretic—there still remains to be explained how the Greeks, rejecting so extensively and selecting with such fine discrimination, came to retain one of the most barbarous features of Egyptian architecture. If the transmission were indirect; if, in quitting Africa, the system took the Asiatic coasts and islands in its way, then the question derives no illustration from the mere reference to Memphis or Thebes, but requires for its solution a wider range of collocation than can be admitted here.

TEMPLE OF JUPITER OLYMPIUS.

PLATE LIV.

IN "a south-eastern direction from the Acropolis, at the distance of about five hundred yards from the foot of the rock, stand sixteen gigantic columns, of the Corinthian order of architecture. They are the remains of a temple which formerly boasted of an hundred and twenty (124?); so disposed as to present a triple row of ten in each front, and a double row of twenty in the flanks. The length of the temple, measured upon the upper step, was three hundred and fifty-four feet; its breadth, one hundred and seventy-one. The columns of this stupendous edifice were six feet and a half in diameter, and more than sixty feet high. The entire building was constructed with the marble from the quarries of Pentelicus."—Wilkins, *Atheniensia.*

"The cluster of magnificent columns of Pentelic marble, at the south-east end of the city, near the Ilissus, belonged to the temple of Jupiter Olympius.

They are of the Corinthian order, sixteen in number, six feet and a half in diameter, and above sixty feet high, standing upon an artificial platform, supported by a wall, the remains of which show that the entire circuit of the platform was two thousand three hundred feet. It appears from the existing remains that the temple consisted of a cell, surrounded by a peristyle, which had ten columns in front, and twenty on the sides; that the peristyle, being double on the sides, and quadruple at the posticum and pronaos, consisted altogether of one hundred and twenty (?) columns, and that the whole length of the building was three hundred and fifty-four feet, and its breadth one hundred and seventy-one. Such vast dimensions would alone be sufficient to prove these columns to have belonged to that temple, which was the largest ever built in honour of the supreme pagan deity, and one of the four most magnificent ever erected by the ancients."*—(Leake, *Topography of Athens.*)

* We cite from Vitruvius, as translated by Gwilt, that portion of the Proem to his seventh book, where he touches incidentally on this subject.

" This work "— the temple of Jupiter Olympius—" is not only universally esteemed, but is accounted one of the rarest specimens of magnificence. For in four places only are the temples embellished with work in marble, and from that circumstance the places are very celebrated, and their excellence and admirable contrivance is pleasing to the gods themselves. The first is the temple of Diana at Ephesus, of the Ionic order, built by Ctesiphon of Gnosus, and his son Metagenes, afterwards completed by Demetrius, a priest of Diana, and Pæonius, the Ephesian. The second is the temple of Apollo, at

Sir William Gell gives fifty-eight feet as the height of the columns, including the architrave; and describes the exterior columns as having plinths under their bases, but the inner range as standing upon a continuous "step," one foot eight inches in height. The walls of the peribolus are built of materials taken from more ancient edifices, and exhibit many an interesting fragment of antique inscriptions.

The history of this gorgeous structure is too curious to be passed over. That the original Olympieum was one of the most ancient of the Athenian temples, may be inferred from the accredited legend which referred its foundation to Deucalion. Pisistratus, about 530 B.C., projected a new and more magnificent erection, but the names of his architects have been better preserved than the works which they superintended. Vitruvius informs us that "the foundations of the temple of Jupiter Olympius at Athens were prepared by Antistates, Callæschrus, Antimachides, and Porinus, architects employed by

Miletus, also of the Ionic order, built by the above-named Pæonius, and Daphnis, the Milesian. The third is the Doric temple of Ceres and Proserpine, at Eleusis, the cell of which was built by Ictinus, of extraordinary dimensions, for the greater convenience of the sacrifices, and without an exterior colonnade. This structure, when Demetrius Phalereus governed Athens, was turned by Philus into a prostyle temple, with columns in front, and by thus enlarging the vestibule, he not only provided accommodation for the novitiates, but gave great dignity to its appearance. Lastly, in Athens it is said that Cossutius was the architect of the temple of Jupiter Olympius, which was of large dimensions, and of the Corinthian order and proportions."

Pisistratus, after whose death, on account of the troubles which affected the republic, the work was abandoned. About two hundred (350 ?) years afterwards, King Antiochus (Epiphanes) having agreed to supply the money for the work, a Roman citizen, named Cossutius, designed with great skill and taste the cell, the dipteral arrangement of the columns, the cornices, and other ornaments. This work is not only universally esteemed, but is accounted one of the rarest specimens of magnificence."—(*Gwilt's translation.*) Notwithstanding this promising commencement, the work was again interrupted by the death of the Syrian monarch; and after the lapse of nearly a century Sylla laid unceremonious hands upon the columns of the unfinished building, and carried them off to Rome, for the purpose of ornamenting the temple of Capitoline Jove. In the time of Augustus, a sort of joint-stock company of kings, states, and wealthy individuals, undertook the completion of the building; but the spell was not yet broken, and the work remained unfinished until the munificence of Hadrian, under happier auspices, finished and dedicated the temple, and set up in it the statue of the god, nearly seven centuries after its foundation by Pisistratus.

It is to be regretted that these interesting ruins have not yet been thoroughly examined. The height of the columns prevented Stuart from ascertaining the details of the capital and entablature; and it was not until 1820 that Mr. Vulliamy succeeded in ascertaining, by means of a projected line and a rope-ladder, the details and dimensions of those members.

"It is hardly possibly to conceive where and how the enormous masses have disappeared of which this temple was built. Its remains are now reduced to a few columns which stand together at the south-east angle of the great platform which was once planted, as it were, by the long files of its pillars. To compare great things with small, they there look like the few remaining chess-men, which are drawn into the corner of a nearly vacant chess-board, at the conclusion of a game."—(Wordsworth, *Athens and Attica.*)

PLATE LIV.

Plan of the temple and peribolus. The darker portions show the still existing fragments.

It must be admitted that these are but meagre particulars of an edifice admired in its own time, not only for its singular magnificence, but as one of the most successful attempts of a later and inferior school, in emulation of the illustrious men who flourished in the palmy days of Greece. The original article in Stuart's great work is brief and unsatisfactory, while his diagrams are few and incomplete. It appears, in fact, that his investigations were so far interrupted, as to leave undetermined the precise length of the temple and the number of columns on its flank. The researches of Mr. Vulliamy have not, that we recollect, been published. In the absence, therefore, of more distinct and technical details, the following fragmentary excerpts from Dodwell may be found to add something to the amount of our information:

"Not a tenth part of the original edifice remains. . . . . . It stands upon a foundation of the soft Piræan stone, like the Parthenon . . . . . . The capitals are not all exactly similar in their ornaments; and are so large that they are composed of two blocks . . . . . The shafts of the columns consist of several frusta . . . . . Part of the peribolus of the Olympieion remains on the south side, facing the Ilissus, and on the eastern end, opposite Hymettus; and a small part of it is visible near the arch of Hadrian; it is composed, in the most perfect part, of eleven layers of stone regularly constructed, and fortified by projecting buttresses, similar to the peribolus of a temple at Delphi."

This instructive, but somewhat indefinite describer, goes on to cite the authority of Pausanias in justification of the statement that the temple was "dipteral and hypæthral." Now Pausanias does not precisely say this, nor is it particularly easy to ascertain what he really does say. It seems never to have occurred to that pleasant but whimsical traveller, that the time might come, and that his sketches *en route* might last until that time, when information such as he could give would be invaluable. He was evidently capable of something better than might be inferred from what he has actually done; and the subjoined extracts, freely rendered from his "Attica," if they do not show what the Olympieum itself was, will, at least, give some idea of its magnificent accompaniments.

"The Emperor Hadrian not only raised the temple of the Olympian Jupiter, but placed in it the colossal and chryselephantine statue of the god, admirable, not only for its magnitude, but for its excellent workmanship. Before you enter the temple, you observe four statues of the Emperor Hadrian,—two in Thasian, and two in Egyptian marble. Before the columns are statues of brass, erected by the colonial cities. The entire peribolus is filled with votive statues, in honour of the Emperor Hadrian. Within the same space are ancient monuments worthy of note; a bronze Jupiter, the temple of Saturn and Rhea, and an enclosed space (τεμενος) dedicated to the Olympian region. There, too, is an opening in the ground, of about a cubit's width, where they say the waters of Deucalion's deluge drained off. Into this chasm are annually thrown cakes of wheaten flour kneaded with honey. There are, moreover, a column on which stands the statue of Isocrates, and a tripod of bronze supported by statues in Phrygian marble."

To these discursive illustrations the critical comment of Lord Aberdeen may supply an appropriate close. "The Roman conquest," he observes, "spread the Corinthian style throughout Greece, almost to the exclusion of the other orders. Although the buildings of this period are often more splendid and costly than those of preceding times, yet the pure taste and correct design of the latter ages of the art are generally wanting. From this remark, however, must be exempted some of the works of Hadrian, the liberal benefactor of Greece; especially if the columns at Athens, which are called by his name, and which are in reality the ruins of the temple of Jupiter Olympius, owe their origin to this emperor. These display the utmost beauty and propriety, with perhaps the greatest degree of magnificence and grandeur, ever attained to by the architectural exertions of the emperors of the Roman world . . . . . . Whenever, or by whomsoever, finished, these columns bear the indications of a pure age of Grecian art; and indeed the remains of such a temple with columns composed of the purest marble, more than six feet and a half in diameter, and sixty feet in height, cannot be described in any terms commensurate with the sensations excited by the view of the original."

THE ARCH OF HADRIAN.

PLATES LV. LVI. LVII. LVIII. LIX.

WO circumstances connected with this structure have furnished materials for a controversy, chiefly remarkable for the readiness with which hypotheses are framed, and the ease with which they are demolished. The first difficulty that presents itself to the archaiologist, when endeavouring to ascertain the object of this insulated monument, is suggested by its very singular position, awkwardly close to the precinct of the Olympieum, and forming an angle with it of very considerable obliquity. Sir William Gell, indeed, states that this arch is actually included within the line of the peribolus; but that this is not the fact is put in the clearest evidence by other authorities, and may be illustrated by a reference to the preceding plate (liv.), where the situation of the two buildings is satisfactorily shown, by the introduction of the monument in its relative position. It stands "within a few yards of the north-east angle of the peribolus;" and the complete finish of all its parts makes it evident that it was never attached to any other construction, while the absence of every mark that might indicate the former existence of a door or gate, proves that it was simply an ornamental erection, probably standing athwart the street leading from the new Agora to the temple of Jupiter, and having its obliquity to the latter determined by the direction of the road. It is further suggested by Colonel Leake, that such a position was in strict accordance with the practice of the Greeks, who usually managed the line of access to their temples, so as to throw them into angular perspective, exhibiting at once the front and flank to the eye of the approaching observer.

A second difficulty has been raised by a sufficiently strange accommodation of certain inscriptions on the entablature, and has afforded opportunity for sundry untenable speculations touching the topography of Athens. Into these, however, it is not expedient to enter here; the discussion would answer no purpose of illustration, to say nothing of its interference with the proper object of a work which professes to deal only with facts and their explanations.

It seems not improbable that the arch of Hadrian was built on the site of a more ancient building, known as the gate of Egeus, though altogether on a different plan. The order is Corinthian, and the material is the Pentelic marble: no cement was used in the construction, but the blocks were held together by metal cramps. Both fronts are precisely similar.

PLATE LV.

Fig. 1. Ground-plan of the arch of Hadrian. The pedestals traced in outline are not now in existence, and the columns have been removed from their places both in front and rear.

Fig. 2. Plan of the upper range: the portions still existing are in shade.

Fig. 3. Section of the building, through the centre of the arch.

PLATE LVI.

Elevation, restored, of the south-eastern front.

PLATE LVII.

Entablature, capital, and base of the antæ of the lower range.

PLATE LVIII.

Entablature, capital, and base of the upper columns.

PLATE LIX.

Fig. 1. Capital and base of the pilasters of the upper range.

Fig. 2. Section of the above capital.

Fig. 3. Section of the soffit and lacunaria under the pediment. It will be observed by an inspection of the plan (Plate lv. fig. 2), that this central compartment is divided by a thin slab of marble into two recesses, or niches.

Fig. 4. Fragment of the ornament which crowns the apex of the pediment.

In the editorial notes to the new edition of Stuart, there occurs a brief and instructive criticism on these details, which shall be given in Mr. Kinnaird's own words, and which could not be inserted anywhere so appropriately as in this place. "It will be observed," he says, "that the antæ at this building have a very sensible diminution, while in structures of the age of Pericles they were never perceptibly diminished. The abaci of all the capitals are painted at the angles, as are those of the Olympieum, of the Pantheon of Hadrian, and of the Incantada at Salonica. No inference as to style, however, can be drawn from this circumstance, as the abaci of the Corinthian capitals at the ancient temples of Phigalia and Apollo Didymæus appear to have been also painted. The introduction of the Ionic echini beneath the foliage of the capitals of the antæ, savours of the declension of pure Grecian art."

It is hardly within the range of this matter-of-fact manual to venture on general criticism. It may, however, be permitted us to hazard the opinion, that this ornamental structure does but add one to the already numerous instances that might be produced, to illustrate the impracticability of harmonising the arch and impost with the column and epistylia. The attempt to combine them with good effect has failed in every instance—in the colossal magnificence of the Flavian Amphitheatre—in the crowded decorations of the triumphal arch—in the whimsical assemblage of discordant features which distinguishes the "Arch of Hadrian." We admit, however, that in the present instance, the principle is not fairly brought to the test: the archivolt breaking in upon the entablature, and the insignificant pediment perched on the centre of the upper story, are inexcusable faults, at variance with all sound theory, and destructive of all legitimate effect.

THE AQUEDUCT OF HADRIAN.

PLATES LX. LXI.

FROM the examples of Greece and Rome, it might be proved clearly enough that magnificence in public structures does not necessarily imply convenience in common domestic architecture. It is equally remarkable that one of the least dispensable sources of health and comfort, a free command of good water, appears to have been deficient in many of the cities of Greece. Even in the fortresses, the wells or springs seem to have been generally without the walls; nor does it appear that tanks were constructed within the circuit of the fortifications. The ornamental edifice now under consideration is a proof both of a scanty supply of potable water, and of the liberality exercised by the munificent Hadrian towards his good city of Athens. Commencing an aqueduct at Cephissia, a "delightful village," even now abounding in springs, he brought a copious stream of wholesome water from a distance of nearly seven miles, to a capacious receptacle at the foot of Mount Anchesmus, and of this reservoir the Ionic building exhibited at Plate lx. in a state of partial restoration, was probably the most conspicuous decoration. A brief inscription on the frieze and architrave recorded the commencement of the undertaking by Hadrian, and its completion by Antoninus Pius.

This building no longer exists, even as a ruin, nor is its site to be traced without difficulty.

PLATE LX.

Plan, elevation, and section, of the "frontispiece." The arch, instead of supporting the entablature, divides it; and the archivolt, instead of springing from a regular impost, or from the cymatium of the frieze, rests immediately on the architrave.

PLATE LXI.

Base, capital, and entablature. These have more of the Roman than the Greek character; yet, though not exhibiting the exquisite finish of the Periclean structures, they are probably superior to any of the existing examples in Rome itself.

THE MONUMENT OF PHILOPAPPUS.

PLATES LXII. LXIII. LXIV.

HIS curious relic of antiquity crowns the summit of the Museum, a considerable elevation, lying to the south-west of the Acropolis. The entire construction is crowded and complicated, but the general form is simple, and ingeniously enough adapted to the advantageous display of sculpture, evidently the great purpose of the erection. The main edifice consists of a mass of masonry, presenting a semicircular front, profusely decorated with pilasters, entablature, niches, statues, and groups in relief. There may be, in all this, somewhat of over-doing, not in strict accordance with the pure and perfect taste of earlier times; yet is the effect rich and impressive, and a partial alteration in some of the lines, with a little abatement of the prevalent redundancy, might have made of this magnificent monument a model for similar structures. It consists of a pilastrade, supported by a basement; and the divisions thus formed are made the framework of three niches, each containing, when entire, a sitting statue; and of three recesses, exhibiting in a continued series of grouped figures the triumphal procession of a Roman emperor. This distribution will be more clearly understood by a reference to the sixty-third plate, where the vacant spaces indicate the disappearance of one entire flank of the building.

It appears, from inscriptions, that this monument was erected by a descendant of the kings of Syria, Caius Julius Antiochus Philopappus, who, in the decay of the fortunes of his family, had been patronised by Trajan, and after attaining the Consular dignity, retired to Greece, and enrolled himself among the citizens of Athens. That this mausoleum was set up in his own lifetime is not, however, absolutely certain; and it may be the more probable opinion that his kindred raised it, as a lasting memorial of a character eminent for moral and intellectual qualities. The central statue is that of Philopappus; the two other figures represented the first and the last monarchs of the royal dynasty from which he was descended. The sculptures on the basement were in honour of the Emperor Trajan. The monument, of Pentelic marble, stands on a platform of Piræan stone.

PLATE LXII.

Fig. 1. Plan of the basement.

Fig. 2. Plan of the upper story, showing the form and arrangement of the niches.

Fig. 3. Perpendicular section, through the centre of the principal niche.

PLATE LXIII.

Elevation of the front.

PLATE LXIV.

Capital and entablature of one of the pilasters.

TEMPLE AT CORINTH.

PLATE LXV.

LITTLE is known, and probably little knowledge is attainable, of these most interesting remains. The description in Stuart is exceedingly meagre, and he does not appear to have had opportunity for minute examination. Even by what name to designate the temple is mere matter of guess; and if a judgment may be formed from the attempts in this way that have been made up to the present time, much is not likely to be done by guessing. Something may, however, be reasonably hoped from that more leisurely and accurate survey which Stuart and Revett were prevented from carrying into effect, and which is peculiarly demanded by the circumstances of this impressive ruin. All that can at present be taken for granted seems to be, that it is the most ancient existing specimen of the Doric order : its massive proportions, the simplicity of its forms, the character of its workmanship, and the coarseness of the material, are sufficient indications of its antiquity. Of the twelve columns marked in the plan, seven only now remain, the remainder having suited the convenience of some Turkish governor, who broke them into fragments, and worked them up in the walls of his own house. It is somewhat singular that at Corinth no specimen should occur of the order to which that once flourishing city gave a name; nor has it been observed, that the acanthus is to be found among the plants of the vicinity.

PLATE LXV.

Exhibits the plan and flank elevation of the temple as it stood in the time of Stuart.

RUIN AT SALONICA,

USUALLY CALLED

THE INCANTADA.

PLATES LXVI. LXVII.

REMARKABLE structure, of which the history and the object are alike unknown. It stands in the Jews' quarter of the city, and, probably from some superstitious fancy, has obtained from the local residents, who are of Spanish descent, the current designation of *las Incantadas* — "the Enchanted Figures." That the most learned antiquarians have failed in ascertaining the real character of this building, may be inferred from the marked variation in their speculations on that subject. A triumphal monument — the entrance to a theatre — the propylæa of a forum — have been suggested and rejected in their turns. Mr. Kinnaird, in a very instructive note to the new edition of Stuart, has directed attention to a similar edifice formerly existing in the city of Bordeaux, and barbarously destroyed, so late as the reign of Louis XIV., by the orders of Vauban, to make way for the bastions and ravelins of a new fortification. This is quite enough to awaken the indignation of artists and antiquarians; yet in this case an engineer might plead the stern dictates of official duty: but what can be urged in extenuation of the gross negligence, on the part both of government and of the local authorities, which could suffer a building, so "rich and rare," to be razed to the foundation, without securing draughts and measurements of all its details? Neither can architects or lovers of art be acquitted of negligence, since a comparatively slight effort on their part might have effected the object, and laid posterity under lasting obligations. Perrault has done most in this way, and from his representations a sufficient idea of the general form and construction may be obtained, to show that the character, and probably the uses of the building, were similar to those of the Incantadas. The high authorities of Perrault and Durand (*Parallèle des Edifices*) are in favour of classing these structures with Basilicæ, and this may be considered as the most probable conjecture, though Mr. Kinnaird is inclined to the opinion that this relic of ancient Thessalonica was a "sepulchral monument." The French ruin was known on the spot as "*les Tutelles,*" or "*le Palais Tutele,*" a designation from which no decided inference can be made. A similar building is said by Vinet to have been, toward the close of the sixteenth century, still in existence at Evora, in Portugal.

The Greeks of Salonica solve the puzzling question concerning the origin of this monument, by a tale of enchantment, mixed up with an amour between Alexander the Great and a queen of Thrace. In this whimsical legend, Aristotle is made to enact the part of a victorious conjurer, confounding the vindictive machinations of the injured husband, and making the spells of a rival magician "return to plague the inventor."

A glance at the plate will show that this monument, as now standing, consists of a Corinthian colonnade, with an entablature, supporting a double row of figures attached to the front and rear of a series of short square pillars, on which rests a sort of architrave, crowned by a cornice. The columns stand on pedestals; the sculpture which adorns the attic is in high relief, and of execution varying from "inferior" to "masterly." The personation of the figures is uncertain.

PLATE LXVI.

Plan, elevation, and section, of the Incantada.

PLATE LXVII.

Entablature and capital of the colonnade. The frieze offers peculiarities which ought not to be overlooked; it is ornamented with fluting, and the profile is carved.

It is probable that further examination, both direct and comparative, might place in clearer light the character and object of this building. One instance of comparison has been cited, and another, less specific indeed, but perhaps not the less important, may be found in the recently explored Temple of Jupiter Olympius at Agrigentum, which displays, though in a different style, and far more in the spirit of antiquity, the same marking feature. But little was known of this colossal structure, till the researches of Mr. Cockerell made its details and proportions familiar to the student of architecture, and exhibited a new illustration of the ingenious artifices and consummate skill in adaptation exercised by the Greek builders. The wealth and splendour of Agrigentum may be inferred from the impressive ruins which yet crown the beautiful eminences whence the ancient city looked forth on the richly cultivated plain beneath, and the broad ocean beyond.* The temple of Jupiter, or of "the

* See the admirably selected view of this striking scenery in the "Magna Græcia" of Wilkins.

Giants," as it is popularly designated, was, when entire, distinguished by several remarkable peculiarities. The interior partitions presented a range of lofty and massive pilasters, supporting a heavy and continued entablature, upon which stood a series of gigantic Telamones, supporting on their heads and bent arms an upper and broken epistylium and cornice.* In these figures there was no variation: they were, unlike those of the Incantada, but in strict conformity with Greek principle, repetitions of each other. It is no improbable conjecture that these Telamones or Atlantes, or by whatsoever epithet they may be distinguished, were intended to represent the rebellious and conquered Titans, thus condemned to emblematic servitude in the fane of the victorious god. The temple itself has, for ages, served as an almost inexhaustible quarry of materials for modern buildings, and its few remaining ruins lie scattered about, or confusedly heaped together; among these are to be found the "disjointed parts" of several "giants," but not, as we understand the statement, of one absolutely entire figure: enough, however, was collected to enable the scientific eye of Mr. Cockerell to trace the continuous line that determined the form and altitude. These colossal statues were built in regular courses, exactly following the lines of the masonry against

* It does not appear that the precise character of this member, or its connexion with the roof, has been ascertained. The temple was probably hypæthral.

which they were placed. "The head alone is 3′ 10″ high, and 3′ 0″ wide; the chest is upwards of six feet across; and the whole height could not have been less than twenty-five feet. . . . The sculpture resembles the archaic manner, observed in the Eginetan statues and those works commonly called Etruscan. Their forms are angular and energetic; and seem to be better suited to the architectural purposes to which they are applied, than the more elegant forms of which the pediments are composed."

The temple itself, contrary to the usual practice in the arrangement of Greek architecture, was divided longitudinally into three distinct enclosures, the intervals between the pilasters as well as those between the exterior columns being walled up; a precaution apparently suggested by the loose texture of the stone, and the consequent difficulty of managing it in sufficiently large masses to form epistylia of the requisite strength. The central inclosure is termed by Mr. Cockerell, the Cella; the side-aisles formed the Porticoes. The whole interior colonnade, if such it may be termed, consisted only of pilasters, while the exterior exhibited a series of engaged columns, of little more than half-diameter, but of extraordinary proportions, each of the fluted recesses having width and depth enough to hold an armed man, and the semi-column measuring thirteen feet in full diameter: "the largest, it is presumed, ever executed. In each of the fronts there were seven columns, and double that number in the flanks, the angles included." . . "The width of the cella between the pilasters is two feet two inches more than the nave of St. Paul's, and the height exceeds it by eighteen feet." From the rocky platform to the central angle of the pediment were, in round numbers, 120 feet; the length was 369; the width, 182.

THE ISLAND OF DELOS.

PLATES LXVIII. LXIX. LXX.

O contrast can be more striking than that which exists between the descriptions of the Sacred Isle in its ancient state, and the miserable aspect of barrenness and dilapidation which it actually presents. Delos, the mythologic birth-place of Apollo and Diana, was enriched by commerce and superstition; temples and consecrated monuments adorned its rugged and unproductive surface; and so awful was its sanctity, that even the Persian ravager respected the hallowed soil. Of all this, nothing now remains but an indiscriminate wreck; what the devastation of barbarian warfare had spared, has been swept away by the rapacity of modern "destructives;" and even the few columns which were standing in the time of Stuart have disappeared, either carried off by Levantine voyagers, or mutilated by Turkish masons for their own wretched purposes.

The only distinct and connected remains that were

observed by Stuart and Revett are represented in the illustrations; subsequent travellers have, however, discovered amid the wreck which covers the island, still more interesting fragments—part of the foot of the Naxian Apollo, and indications of the Mithratic worship, in the ugliest of all imaginable capitals, formed by the head and shoulders of a bull couchant.

PLATE LXVIII.

Fig. 1. Plan and elevation of three Doric columns, assigned by Stuart to the Temple of Apollo. The fluting occupies only a narrow circle at the base and at the necking, and is generally supposed to have been left unfinished: it should seem to be quite as probable that this peculiarity was the result of some fantastic aim at novelty, especially as this is not the only instance of the kind.

Fig. 2. Plan and elevation of the "Portico of Philip," a designation given on the authority of an inscription, but liable to considerable uncertainty respecting its precise definition. Stuart seems to have been highly gratified by the "lightness of its proportions," and by its greater suitableness to "common use." It may, however, be permitted to suggest, that whatever may have been gained in this respect, is at the expense of all that is characteristic in the Doric shaft, and that the straight profile of the echinus is mean and ineffective, while the capital altogether is rendered insignificant by deficiency in size and mass.

PLATE LXIX.

Entablature, capital, and shaft of the Temple of Apollo; with details of the capital, on a larger scale.

PLATE LXX.

Entablature, capital, and shaft of the Portico of Philip; with details of the capital, on a larger scale.

THE PNYX.

F the localities connected with Athenian antiquity, none are more decidedly indicated than this remarkable site. It exhibits nothing of complicated construction: a few lines and masses of rock and masonry, with a little levelling of surface, are all that exist—all, or nearly all, that ever did exist—to mark the spot where assembled the "fierce democratie," and whence the "famous orators" who "wielded at will" the passions and the energies of the fiery multitude, "fulmined over Greece."

Yet there has been a good deal of strange blundering about this place. Spon took these remains for those of the Areopagus; Wheler hesitated between the Areopagus and the Odeum; while Stuart decided for the latter. Yet there are all the features assigned by ancient authors to the Pnyx, and quite inapplicable, without much forced construction, to any other spot.

A reference to Plate xl. will explain the few and simple characteristics of the scene; the semicircular boundary, partly scarped and partly built up; the chord forming an obtuse angle; and the Bema, or square block, insulated from the main rock, and forming the pulpitum, or elevated platform, from which the orators addressed the assembled people.

"This," exclaims Mr. Wordsworth, "was the place provided for the public assemblies at Athens in its most glorious times; and nearly such as it was then, is it seen now. The Athenian orator spoke from a block of bare stone: his audience sat before him on a blank and open field."

Yet was this blank and bare arena surrounded by a scene of unexampled splendour, supplying the orator with subjects of exciting appeal to the pride and passions of his audience.

"Visible behind him,"—we again cite Mr. Wordsworth—"at no great distance, was the scene of Athenian glory, the island of Salamis. Nearer was the Peiræus, with its arsenals lining the shore, and its fleets floating upon its bosom. Before him was the crowded city itself. In the city, immediately before him, was the circle of the Agora, planted with plane-trees, adorned with statues of marble, bronze, and gilded, with painted porticoes, and stately edifices, monuments of Athenian gratitude and glory: a little beyond it was the Areopagus; and, above all, towering to his right, rose the Acropolis itself, faced with its

Propylæa as a frontlet, and surmounted with the Parthenon as a crown."

The Pnyx included an area of more than twelve thousand square yards, and could with ease contain the entire free civic population of Athens.

GLOSSARY OF ARCHITECTURAL TERMS,

WITH

OCCASIONAL REFERENCES TO THE EXAMPLES.

ABACUS—A square or quadrangular member interposed between the capital of a column and its entablature. When a similar member is placed beneath the base of a column, it is called a plinth.—*See pls.* 11, 19, 20, 46.

ABUTMENT—The solid masonry, which resists the lateral pressure of an arch.

ACANTHUS—A plant, of which the leaves form an ornament in the Corinthian capital, and are said to have originally given rise to the order.

ACROTERIA—Bases or low pedestals resting on the angles and vertex of a pediment, and intended for the support of statues, or other ornaments.

AILE or AISLE, the lateral divisions which run parallel with the Naos, or Nave, of a temple interior.

AMPHIPROSTYLE—Having a portico at both extremities.

AMPHITHEATRE—A theatre in which the spectators entirely surround a central arena, or pit, in which the exhibitions take place.

ANDRON—A passage, open space, or court.

ANNULET.—A small flat fillet encircling a column, used either by itself or in connexion with other mouldings. It is used several times repeated under the *ovolo* or *echinus* of the Doric capital.

ANTÆ—Pilasters terminating the side-walls of a temple. —*See pls.* 27, 39, 43.

ANTEFIXÆ—Ornamental blocks, vertically affixed at regular intervals along the eaves of a roof, to cover the joints of the tiles.—*See pl.* 17.

ANTHROPOMORPHIC—From the Greek ἄνθρωπος (anthrōpos), a man or human being, and μορφὴ (morphé), form:—shaped like or bearing the semblance of a human figure.

ANTHROPOSTYLE—From ἄνθρωπος and στύλος (stylos), a column; an anthropomorphic pillar.—*See pl.* 5.

APSE or APSIS—The curved or multangular termination of a cathedral choir, &c.

ARABESQUE—A term strictly applied to the flowing and fanciful ornament, which originated with the Arabian architects, and which prevails in all Mahomedan structures.

ARÆOSTYLE—A species of intercolumniation, to which four diameters are allowed between each column.

ARCHITRAVE—The *epistylium*, or beam, extending from column to column. The lower of the three members forming the entablature of the Greek order. The moulded frame which bounds the sides and head of a door or window-opening.—*See pl.* 24.

ARCHIVOLT—A collection of mouldings on the face of an arch, resting upon the imposts.—*See pl.* 60.

ASTRAGAL—A narrow moulding, the profile of which is semicircular.

Astylar—Columnless, or without columns; the absence of columns, where they might else be supposed to occur.

Atlantes—Male figures or statue-pillars, used in the place of columns; so called by the Greeks, but by the Romans called *Telamones.—See pl.* 67.

Attic—A term commonly applied to constructions resting on the entablature.

Balustrade—A breastwork of *balusters* with a plinth and coping.

Base—A general term for the lowest member of any construction. The base of a column is the portion on which the shaft is placed.

Blocking Course—A plain course of stone, forming a low parapet above the cornice of a building or portico.

Buttress—A projecting pier, to strengthen a wall, or resist the pressure of an arch or vault.

Capital—The head or upper part of a column or pilaster.

Caryatid—A female figure supporting an entablature. *—See pls.* 33, 37, 38.

Cassoon—A sunk panel, or coffer, in a ceiling.

Caulicoli—The twisted stalks or volutes under the flower on the abacus of a Corinthian capital, called also Helices.*—See pl.* 16.

Cavea—*See Coilon.*

Cell or Cella—*See Exhedra* and *Naos.*

Choragus (whence *Choragic*)—A term applied in Athens to those who superintended a musical or theatrical entertainment, and provided a chorus at their own expense.*—See pls.* 12–17, 41, 42.

CHRYSELEPHANTINE—Carved in gold and ivory.

COFFERS—Sunk panels in vaults or domes.

COILON — The area of a theatre.—*See pl.* 40.

COLLARIN, or COLLARINO—The neck or frieze of a Doric (or Tuscan) capital.—*See pl.* 14.

COLUMN — A vertical support, generally including a base, shaft, and capital.

CORINTHIAN — The first order in architecture — the lightest and most ornamental. It has two annular rows of acanthus leaves in the capital, each leaf of the upper row alternating with those of the lower, besides other enrichments. Its invention is ascribed to Callimachus, 540 B.C.—*See pls.* 17, 19, 20, 55–59, 64, 65, 66, 67.

CORNICE — The crowning projection of the entablature.

CORONA — The square projection forming the principal member of a cornice, answering to the face of the overhanging eaves of a roof.

CUPOLA — A spherical roof, rising like an inverted cup over a circular or multangular cell. It is commonly, but incorrectly, styled a *Dome*.

CUSPS — The points of tre-foils, quatre-foils, &c.

CYMATIUM — An undulating moulding.—*See pls.* 11, 60.

CYRTOSTYLE — The semicircular porticos projecting from the transepts of St. Paul's are *cyrtostyle*.

DADO, or DIE — The central bulk of a pedestal, between its base and cornice.

DECASTYLE — A portico or building having ten columns in front.

DENTILS — A series of tooth-like ornaments, common to the Ionic and Corinthian cornices.

DIASTYLE — An intercolumniation of about three diameters between each column.—*See pl.* 30.

DIPTERAL — Having a double range of columns.

DITRIGLYPH—An interval between two columns, admitting two triglyphs on its entablature.

DORIC — The second order in architecture, (somewhat between the Ionic and the Tuscan,) distinguished for simplicity and strength. — *See pls.* 3, 4, 23, *et seq.*, 43, *&c.*, 48, *&c.*, 65, 68, *&c.*

ECHINUS — Properly the egg-and-anchor ornament peculiar to the Ionic capital ; it is sometimes used for the whole moulding instead of ovolo.

ENTABLATURE — The horizontal superstructure of a colonnade ; in Greek architecture, comprising the three divisions of architrave, frieze, and cornice.

ENCARPUS — Festoon of fruit or flowers on friezes, &c.

EPISTYLIUM.—The Architrave ; the lowest of the three divisions of an entablature.—*See pls.* 59, 67.

EUSTYLE—An interval of two diameters and quarter between the columns.

EXHEDRA—A recess, or small side-apartment.—*See pl.* 19.

FAÇADE—Front view or elevation of an edifice.

FASCIA—A flat member in the entablature, representing a band, or broad fillet. The architrave of the more enriched orders is divided longitudinally into two or more fasciæ.

FENESTRATION, termed by the Germans *Fenster-architektur*, is, in contradistinction to columniation, the system of construction and mode of dēsign marked by windows.

FLATBAND—A flat fascia or moulding, of which the projection is less than its breadth; also the lintel of a door or window, or fillet between the flutings of a column.—*See pl.* 34.

FLUTING—The vertical channelling of the shafts of columns.

FRIEZE—The middle member of an entablature, frequently decorated with raised carvings.

FUST—The shaft of a column, or trunk of a pilaster; *i.e.* that part which is between the base and capital.

GLYPHS—The perpendicular channels cut in the triglyphs of the Doric frieze.

GUTTÆ, or drops in the Doric entablature, are small pyramids or cones, immediately under the triglyph and mutule.

HELICES—*See Caulicoli.*

HERMES—A square shaft or pillar terminating in a bust.

HEXASTYLE—Having a front range of six columns.—*See pls.* 29, 33.

HYPÆTHRAL—Without a roof.

HYPERTHYRUM—The upper member of a doorway.—*See pl.* 32 *bis.*

HYPOTRACHELIUM—The necking of a capital, introduced between the capital itself and the shaft of the column.

IMPOST—The member on which the arch immediately rests.—*See pl.* 60.

INTERCOLUMNIATION—The space between two columns.

IONIC—The third order in architecture, founded by the Ionians about 1350 B.C. Its distinguishing characteristics are the slenderness and flutings of its column (although less slender than the Corinthian), and the volutes of rams' horns which adorn the capital.—*See pls.* 5, *&c.*, 30, *&c.*, 41, 61.

JAMBS—The side-supports of a doorway.

KEYSTONE—The top stone or central wedge in the curve of an arch.

LACUNARIA—The sunk panels or coffers in ceilings.

LINTEL.—The part of a door or window-frame that rests on the side-posts.—*See pl.* 47.

METOPE—The interval between the Doric triglyphs.—*See pl.* 48.

MODILLION—An ornament resembling a bracket in the Corinthian (and Composite) cornices.

MODULE—The semi-diameter of a column, or thirty minutes.

MONOTRIGLYPHIC—That mode of intercolumniation in the Doric Order, according to which there is only a single triglyph over each intercolumn.

MUTULES—Plain projecting blocks, supporting the corona of the *Doric* cornice, answering to modillions in the *Corinthian.*

NAOS—The cell or central chamber of a temple.

NECKING—The space between the astragal of the shaft and the annulet of the capital.

OCTASTYLE—An edifice having eight columns in front.

ODEUM—A structure built by Pericles at Athens for the performance of music. It had within many rows of seats and of pillars, and the roof was conical.

OPISTHODOMUS—The chamber behind the cella.—*See pl.* 25.

ORDER—In architecture a column entire, consisting of base, shaft, and capital with an entablature.

OVOLO—A moulding which projects one quarter of a circle, called also a quarter-round.

PALÆSTRA or GYMNASIUM—A Grecian structure, in its use answering nearly to the baths.

PEDIMENT—The triangular termination of the roof of a temple, resting upon the entablature which surrounds the building.

PERIBOLUS—The enclosure within which a temple stood. —*See pls.* 19, 54.

PERIPTERY (whence PERIPTERAL)—An edifice surrounded by a range of columns or colonnade.—*See pl.* 48, *&c.*

PERISTYLE—A range of columns within a court or temple.—*See pl.* 23.

PILASTER—An *anta,* or square pillar, generally built, as it were, into the wall, from which it projects one-fifth or one-sixth of its breadth; but sometimes insulated. It has the same proportions and ornaments as a column, but no diminution.

PILASTRADE—A row of pillars.—*See pl.* 63.

PLINTH—The low square step on which a column is placed.

PODIUM—A dwarf pedestal wall.

POLYSTYLE—Having a number of columns. Where columns occur behind columns, as where a portico has inner columns.

PORTICO—The covered space in front of a temple.

POSTICUM—The covered space behind a temple.—*See pls.* 27, 43, 48.

PRÆCINCTIONS—The landings or gangways which separated and gave access to the ranges of seats in the ancient theatres.

PROFILE—The outline of a series of mouldings, or of any other parts, as shown by a section through them.

PRONAOS—Often used convertibly with Portico. The part of the temple in front of the Naos.—*See pls.* 29, 48.

PROPYLÆUM—The piece of advancing architecture which distinguishes the entrance into the court of a temple.

PROSCENIUM.—The place before the scene when the actors appeared; the stage.—*See pl.* 40.

PROSTYLE—A term to distinguish the *open projecting* portico from the portico *in antis;* which latter, having its sides *enclosed* by the continued walls of the cell, appears to have no projection.

PSEUDO-DIPTERAL—False or imperfect dipteral, the inner range of columns being omitted.

PTEROMA—Has frequently a more extended application, but in the present work it is used to distinguish the continuation of the side-walls from the transverse partition of the cella to the antæ.—*See pls.* 19, 43.

PULVINATED—A frieze whose face is convex instead of

plane is said to be *pulvinated*, from its supposed resemblance to the side of a cushion.

PYCNOSTYLE—The first method of intercolumniation, having one diameter and a half between each column.

SCOTIA—From the Greek σκοτία (skotia), darkness: the large hollow or concave surface in the base of a column. The scotia and tori mutually set off each other, as regards light and shade.

SHAFT—The main body of a column between the base and capital.

SOCLE—*See* ZOCLE.

SOFFIT—Ceiling: applied to the underside of arches, and of other architectural members.

SPANDRIL—The space between the curve of an arch and the square enclosing it.

STOA—A portico. In one of these, at Athens, Zeno taught his system of philosophy, and instituted the sect named Stoics.—*See pl.* 18.

STRIGÆ—The flutings of a column.

STYLAGALMATIC—Supported by figure-columns.—*See Caryatid, Atlantes, Telamones.*

STYLOBATE—The basis or substructure on which a colonnade is placed.—*See pls.* 12, 38.

SUPERCOLUMNIATION—The placing of one tier of columns over another.

SYSTYLE—An intercolumniation of two diameters.—*See pl.* 30.

TELAMONES—*See* ATLANTES.

TENIA—The upper member of the Doric architrave.

TERMINUS—A bust surmounting a pedestal, which diminished downwards, and was firmly fixed in the earth, to symbolise the deity which presided over the sacred limits of property.

TETRASTYLE—A building with four columns in front. —*See pls.* 29, 30.

THOLOBATE—Mr. Hosking's term for the substructure of a cupola.

THOLUS—A cupola or circular building: employed in this volume to designate the circular roof of the monument of Lysicrates.—*See pl.* 17.

TOREUTIC—Carving, embossing, working in metals.

TORUS—A large convex moulding or *coil* in the base of a column.

TRIGLYPH—The distinguishing ornament of the Doric entablature.—*See pl.* 18, *&c.*

TYMPANUM—The triangular space within the cornices of a pediment.—*See pl.* 48.

VESTIBULE—A porch or ante-room, through which a larger apartment or house is entered.—*See pls.* 19, 43.

VOLUTE—The Ionic scroll or ram's horn. Those of the Corinthian capital are smaller, and of distinct character.—*See pls.* 35, 36.

VOMITORIA—Passages facilitating egress from a theatre.

XYST—A large court with a portico on three sides, planted with rows of trees, where the ancients performed athletic exercises, as running, wrestling, &c. —*See Palæstra.*

ZOCLE, or SOCLE—A plain low plinth, or pedestal.

ZOPHORUS—The frieze.

ZOTHEKA—A small room or alcove, which might be added to or separated from another, by means of curtains and windows.

CHRONOLOGICAL INDEX.

London:—Printed by G. BARCLAY, Castle St. Leicester Sq.

B

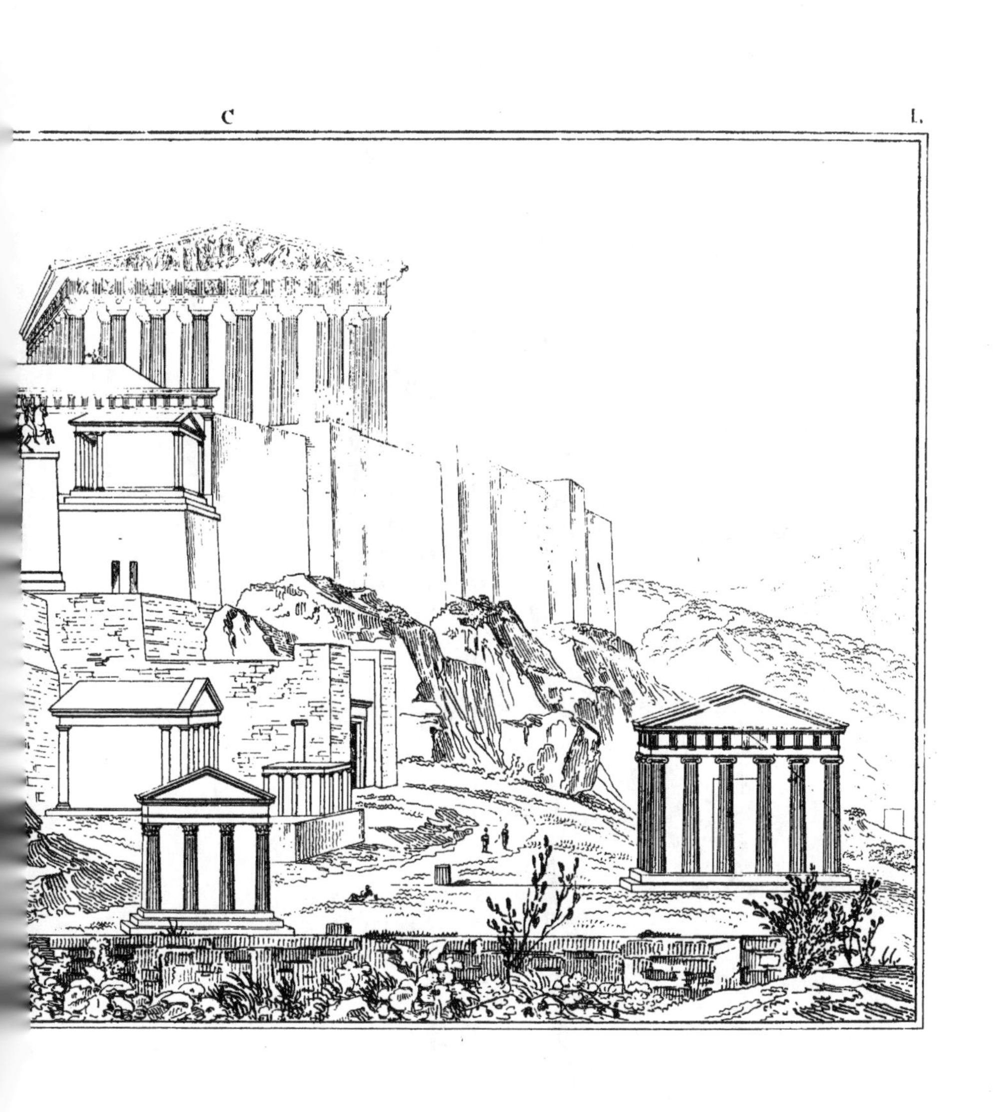
C
L.

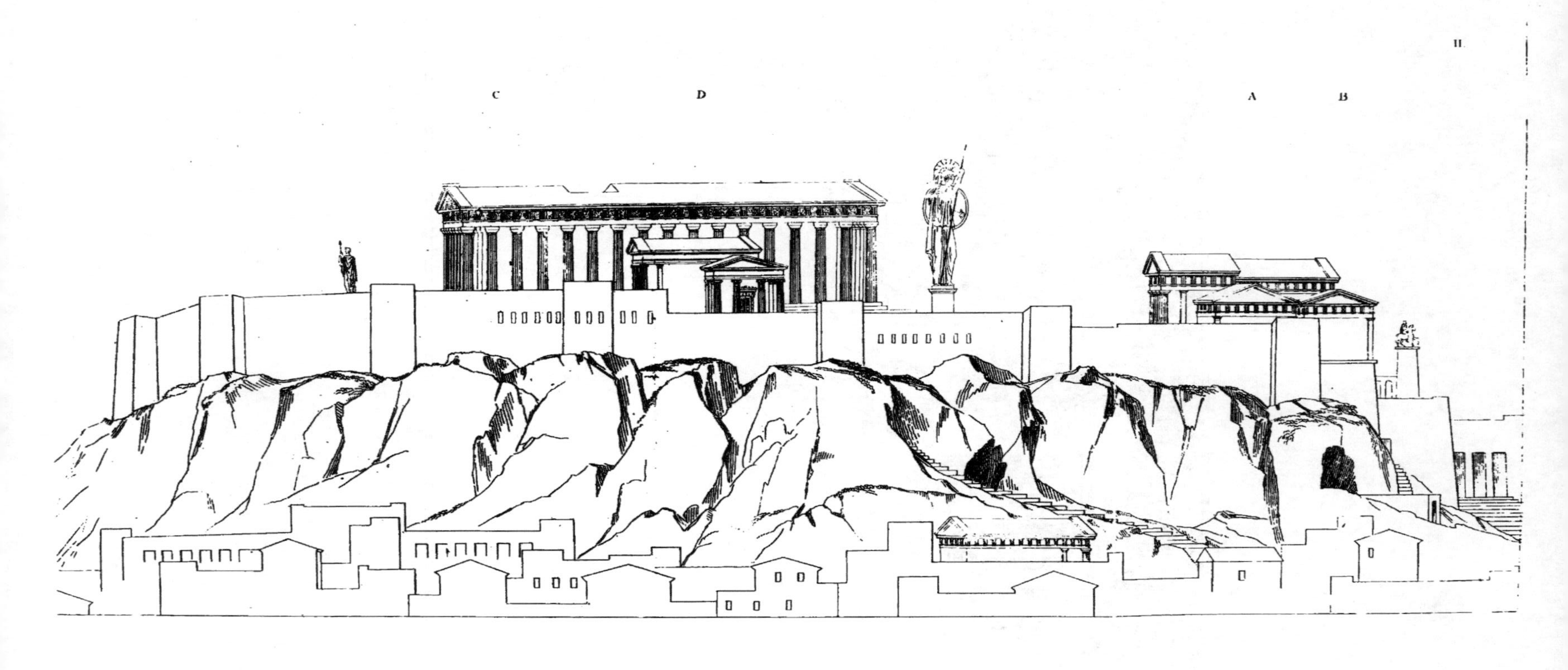
II
C
D
A
B

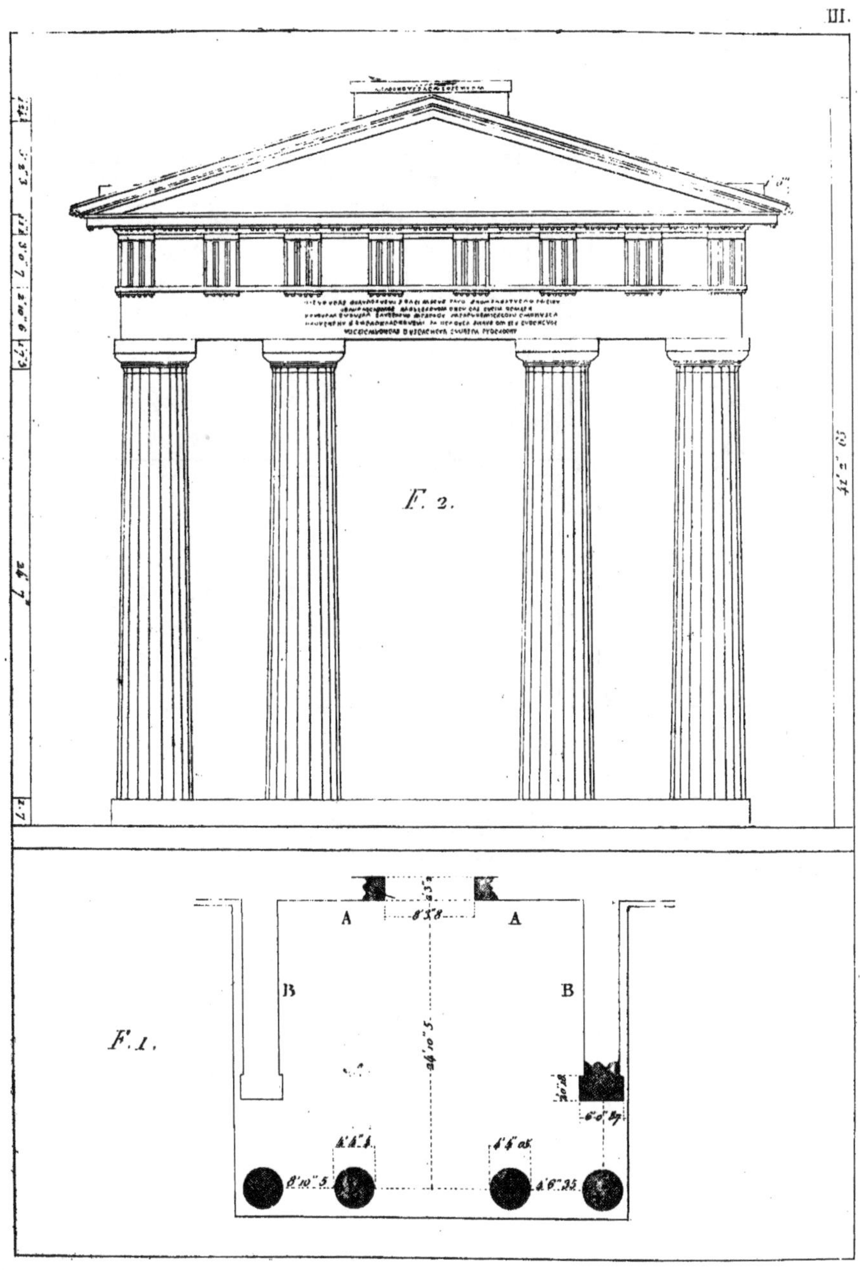
F. 2.
F. 1.
A
A
B
B
8'5"8
24'10"5
8'10"5
4'6"35
4'4"4
4'4"05
6'0"87
42'2"03
26'7"
6
12 mod.

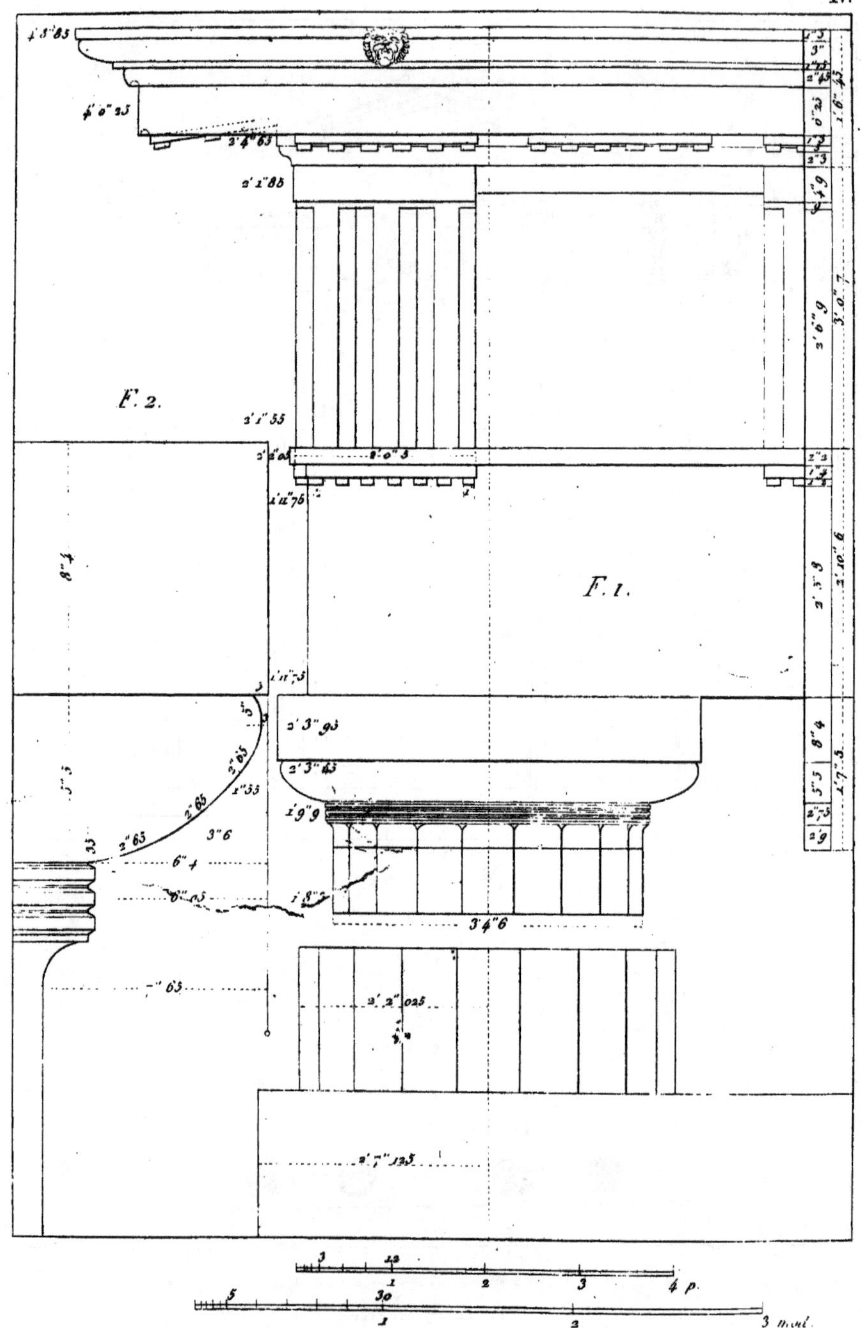
F. 2.
F. 1.
4 p.
3 mod.

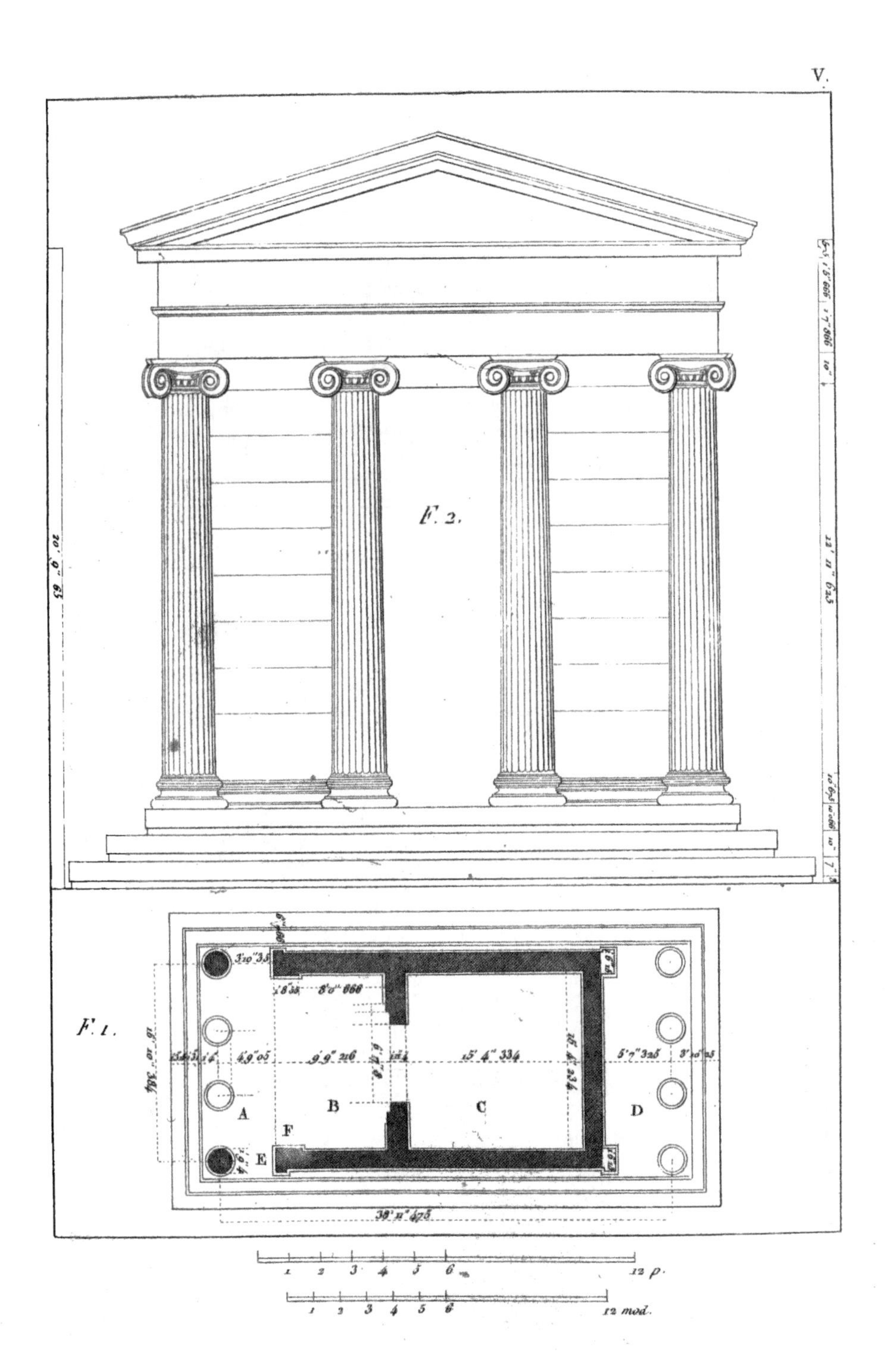
F. 2.
F. 1.
A
B
C
D
E
F
1 2 3 4 5 6 12 p.
1 2 3 4 5 6 12 mod.

1'9"35
1'8"2
1'0"4
10"366
10"366
1'0"2
10"2
10"2
1'0"05
9"1
1'6"2
1'9"4
10"7
11"45
1'0"884
1'1"8
1'2"85
1'3"73
3 p.

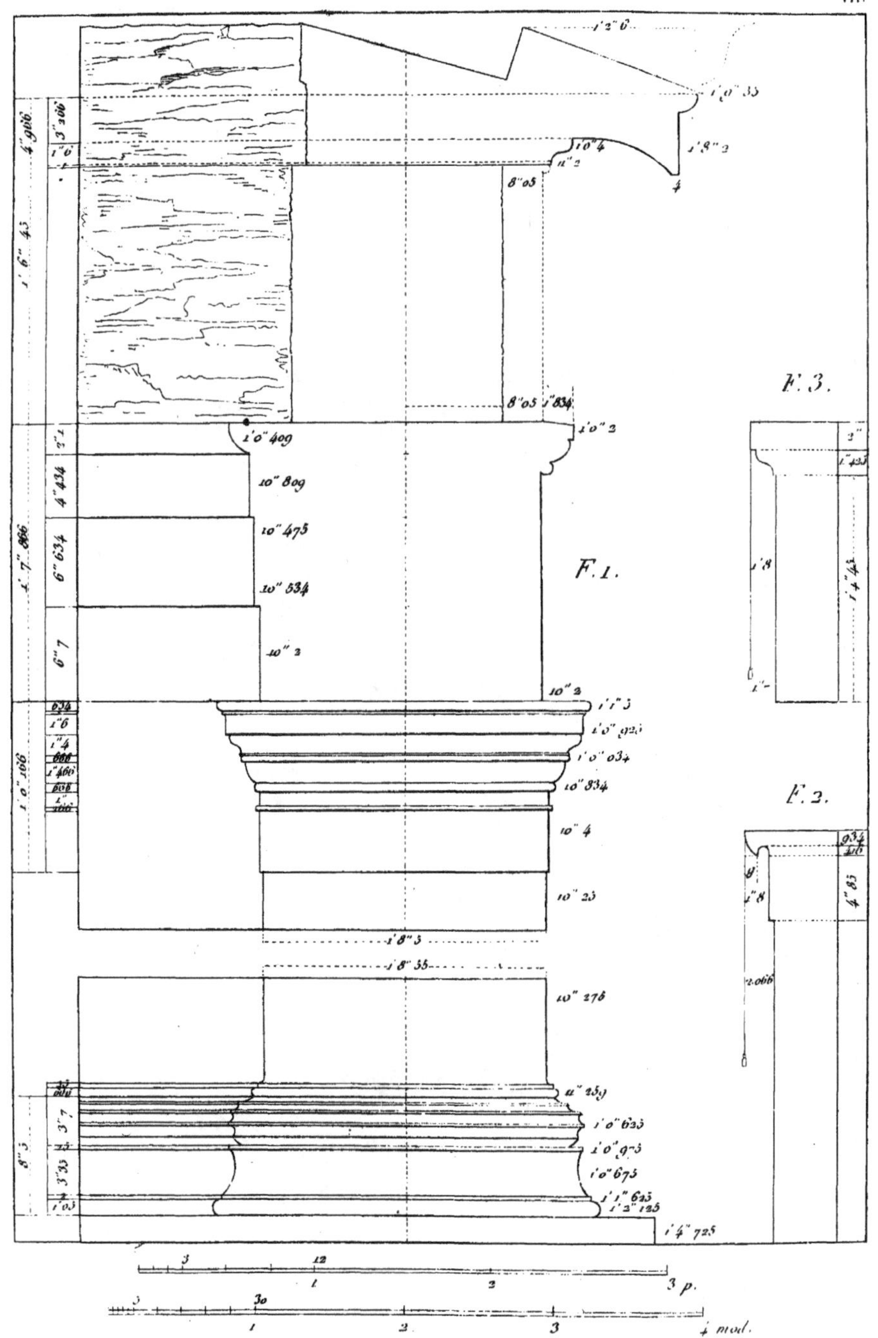
F. 1.
F. 3.
F. 2.
3 p.
4 mod.

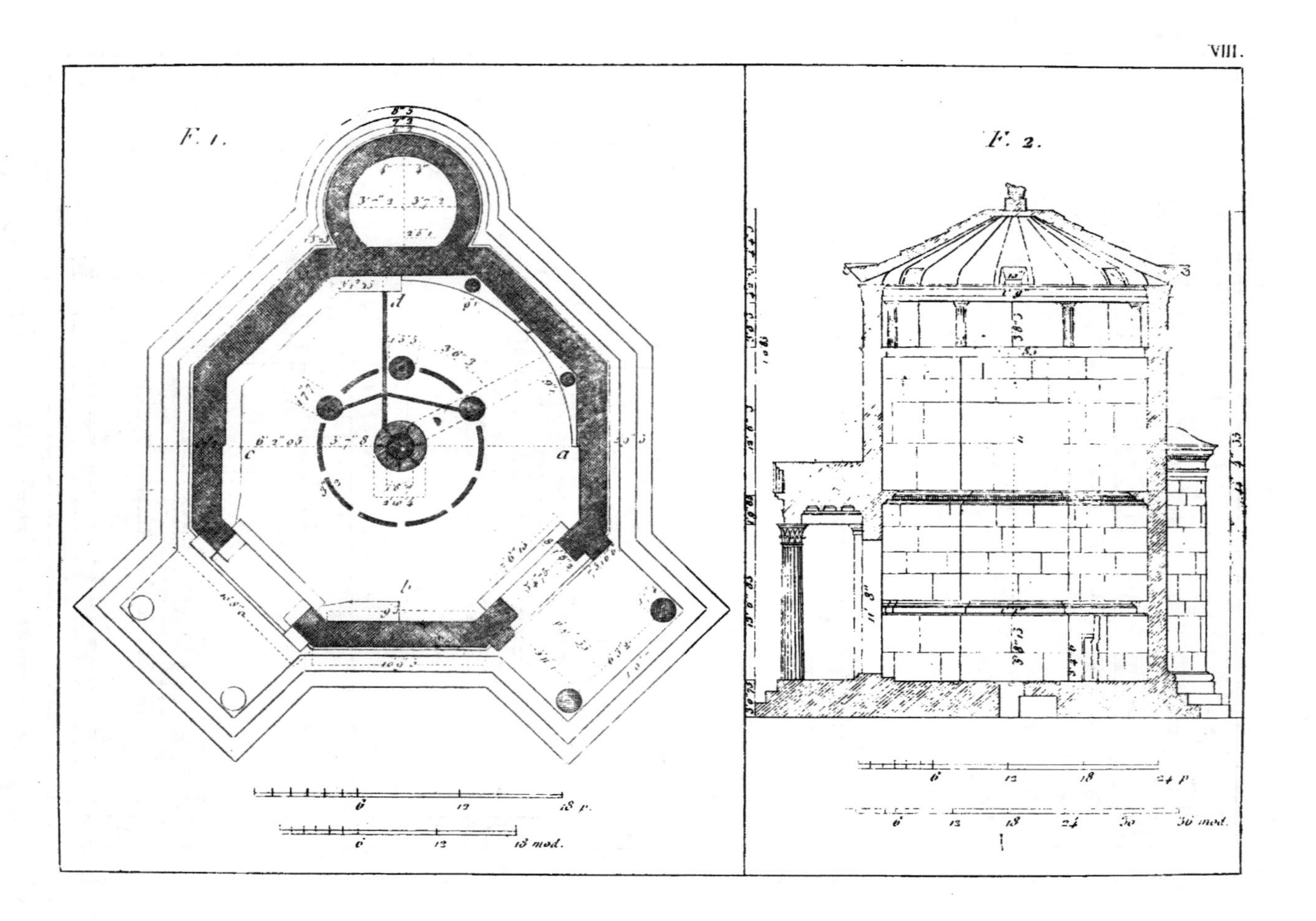
F. 1.
F. 2.
a
b
c
d
6
12
18 p.
6
12
18 mod.
6
12
18
24 p.
6
12
18
24
30
36 mod.

ΑΠΗΛΙΩΤΗΣ ΒΟΡΕΑΣ ΚΑΙΚΙΑΣ

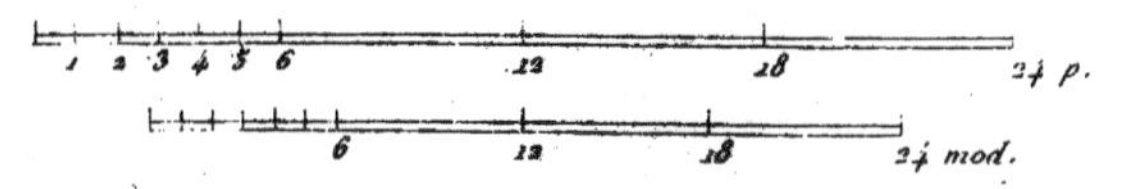

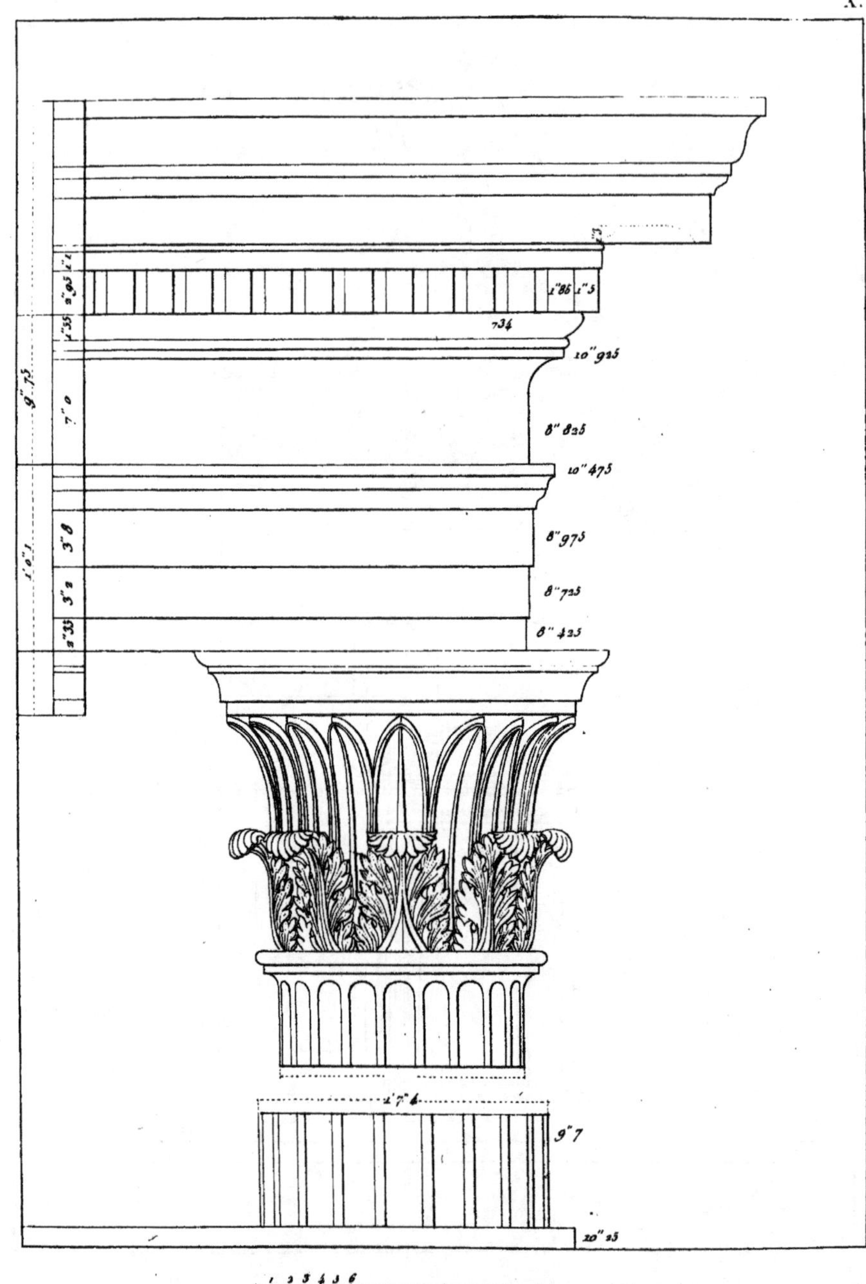

1"86
1"5
734
10"925
8"825
10"475
8"975
8"725
8"425
1'7"4
9"7
10"25

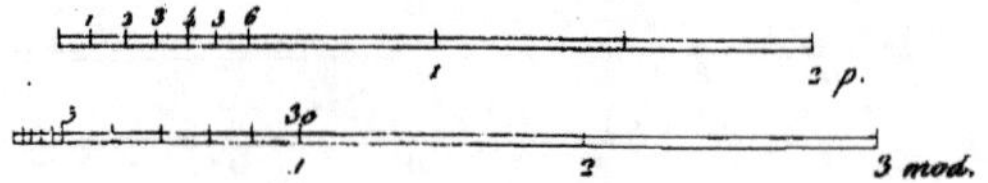

1 2 3 4 5 6
1
2 p.
30
1
2
3 mod.

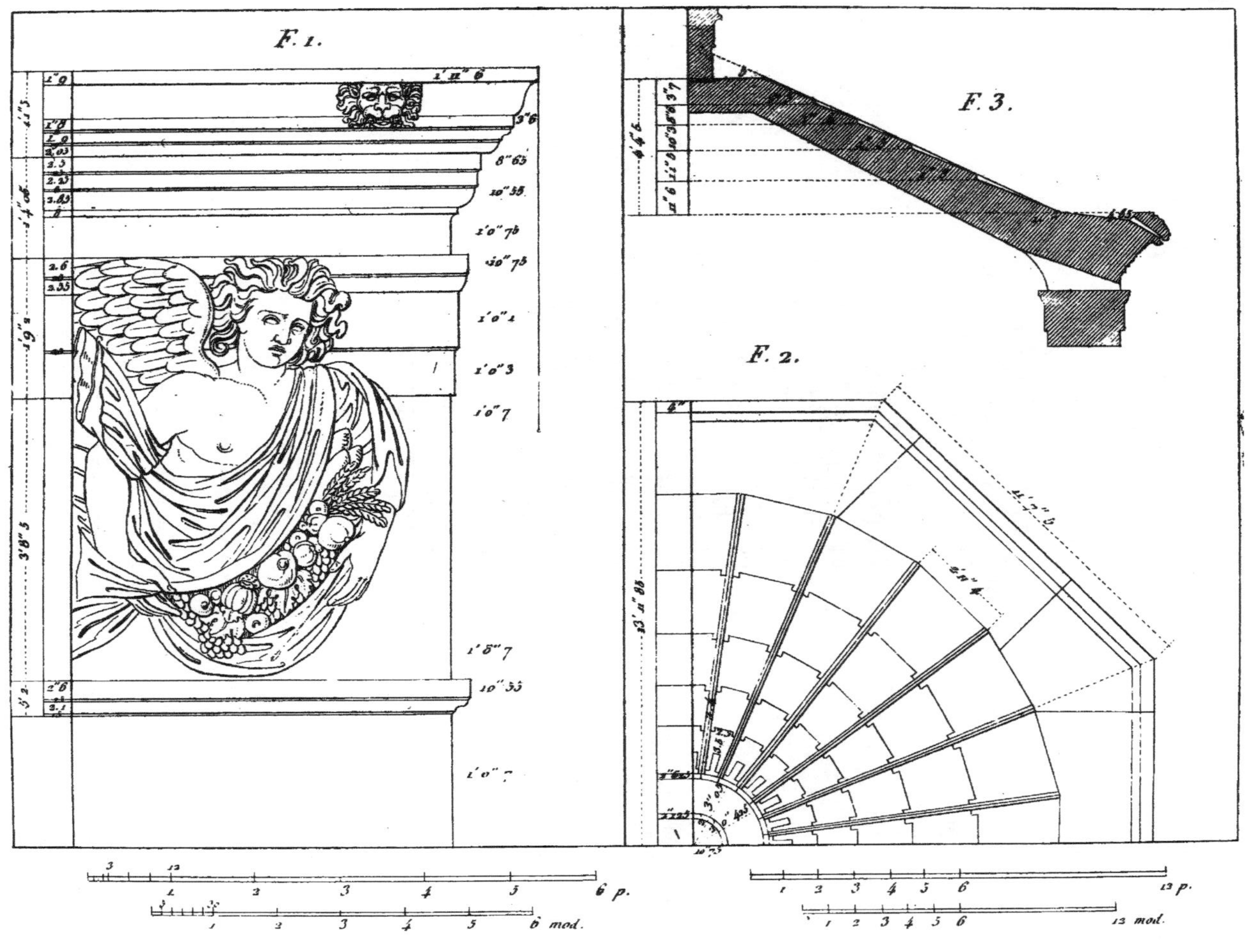

F. 1.

F. 3.

F. 2.

6 p.

6 mod.

13 p.

13 mod.

1 2 3 4 5 6
12 p.
6 12 18
24 mod

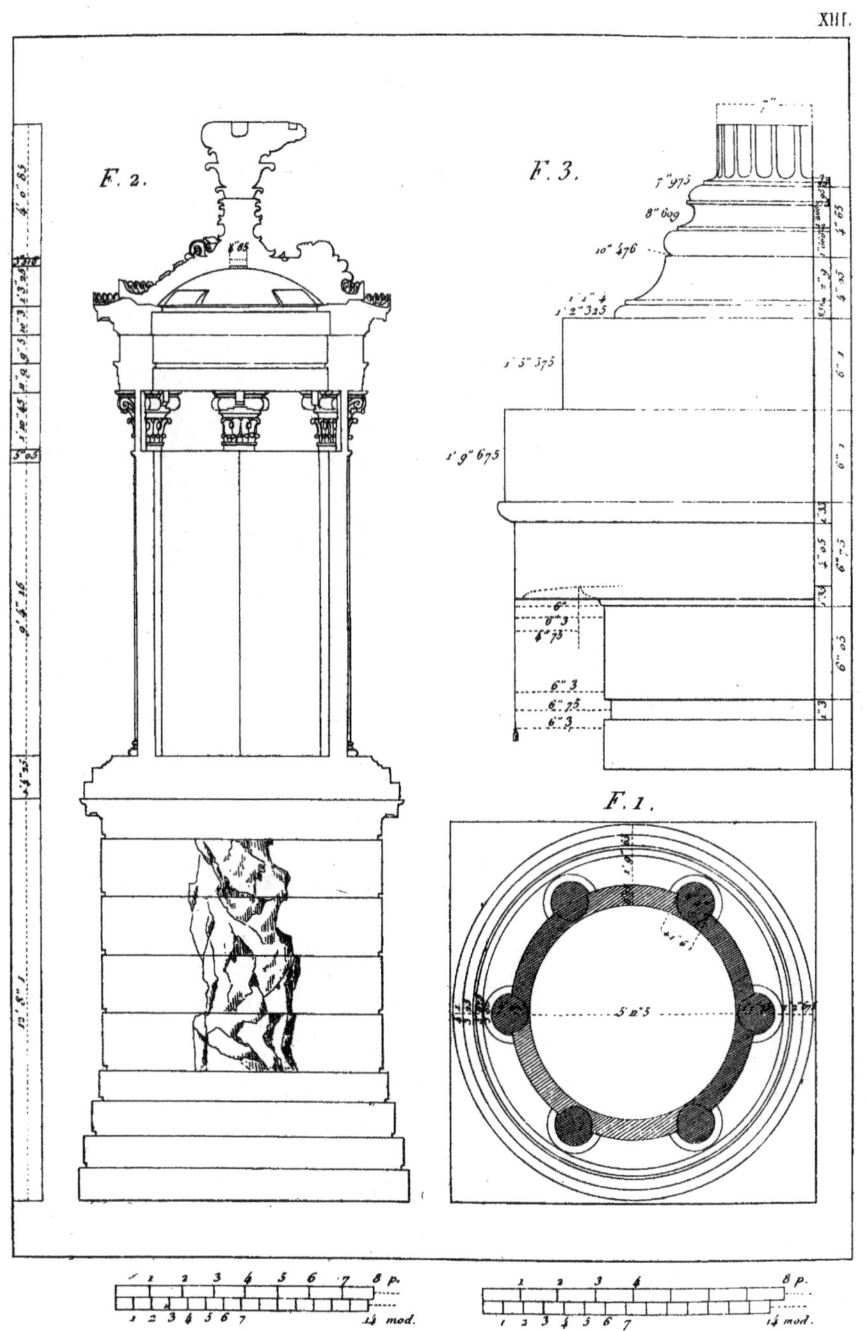
F. 2.
F. 3.
F. 1.
8 p.
14 mod.

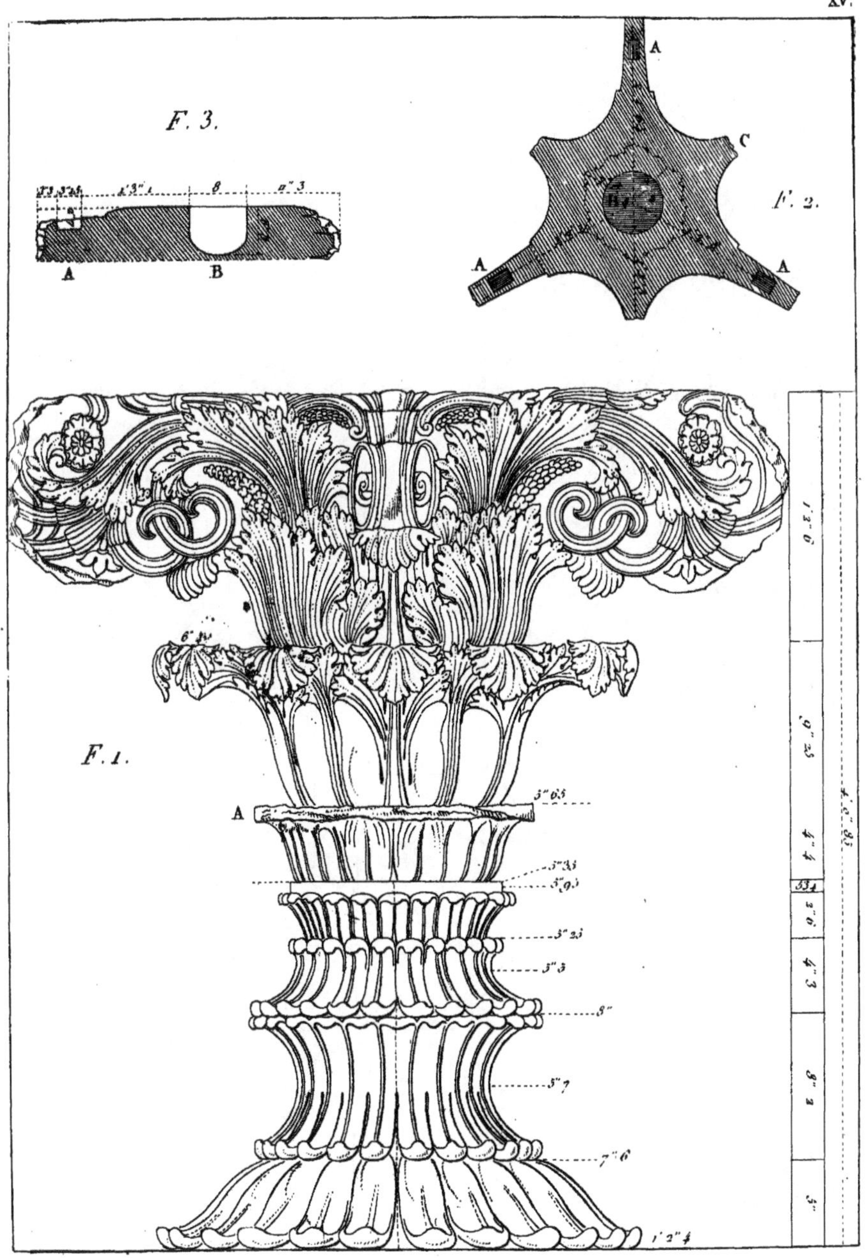
F. 3.
A
B
F. 2.
A
C
A
A
F. 1.
A
1' 2" 4

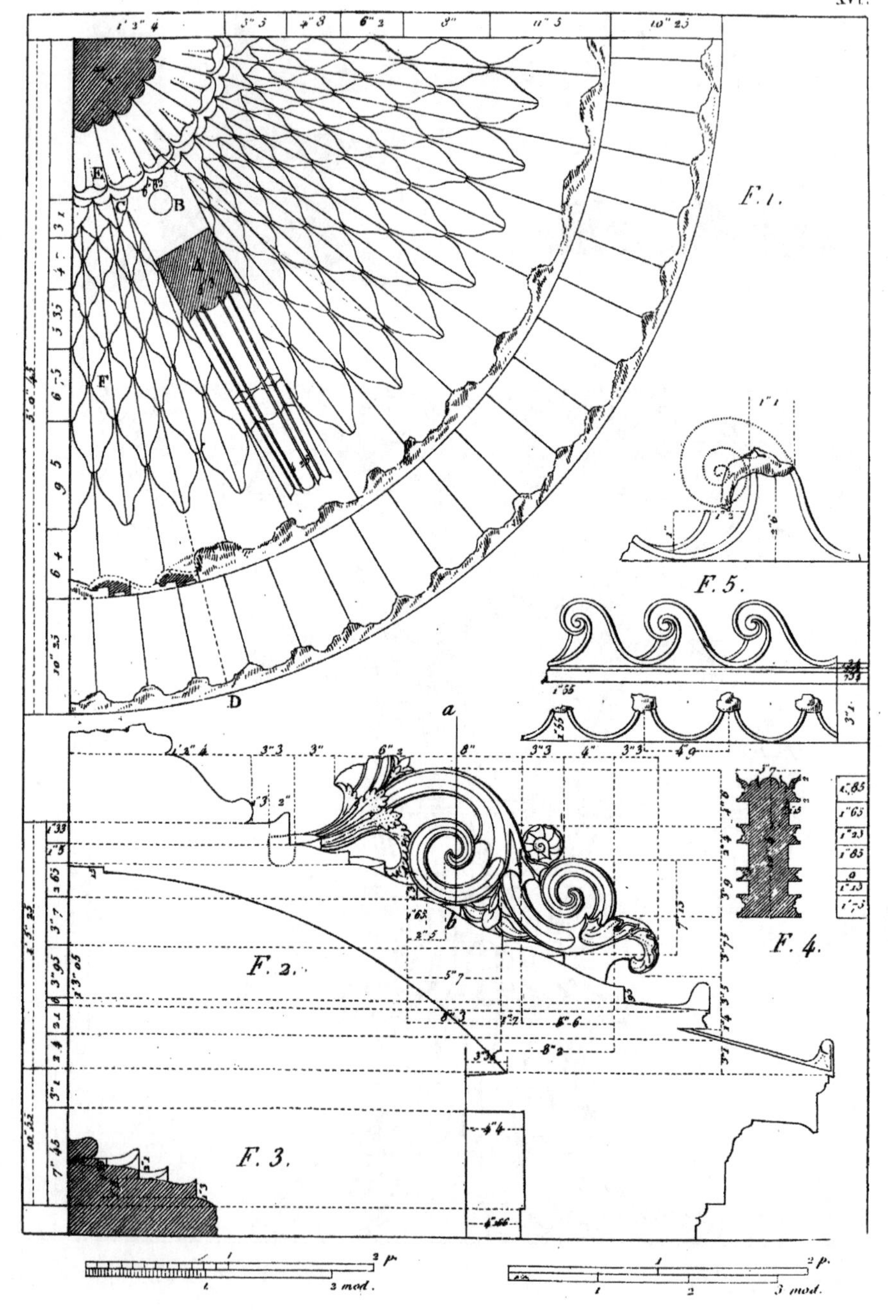
F. 1.
F. 2.
F. 3.
F. 4.
F. 5.
3 mod.
3 mod.

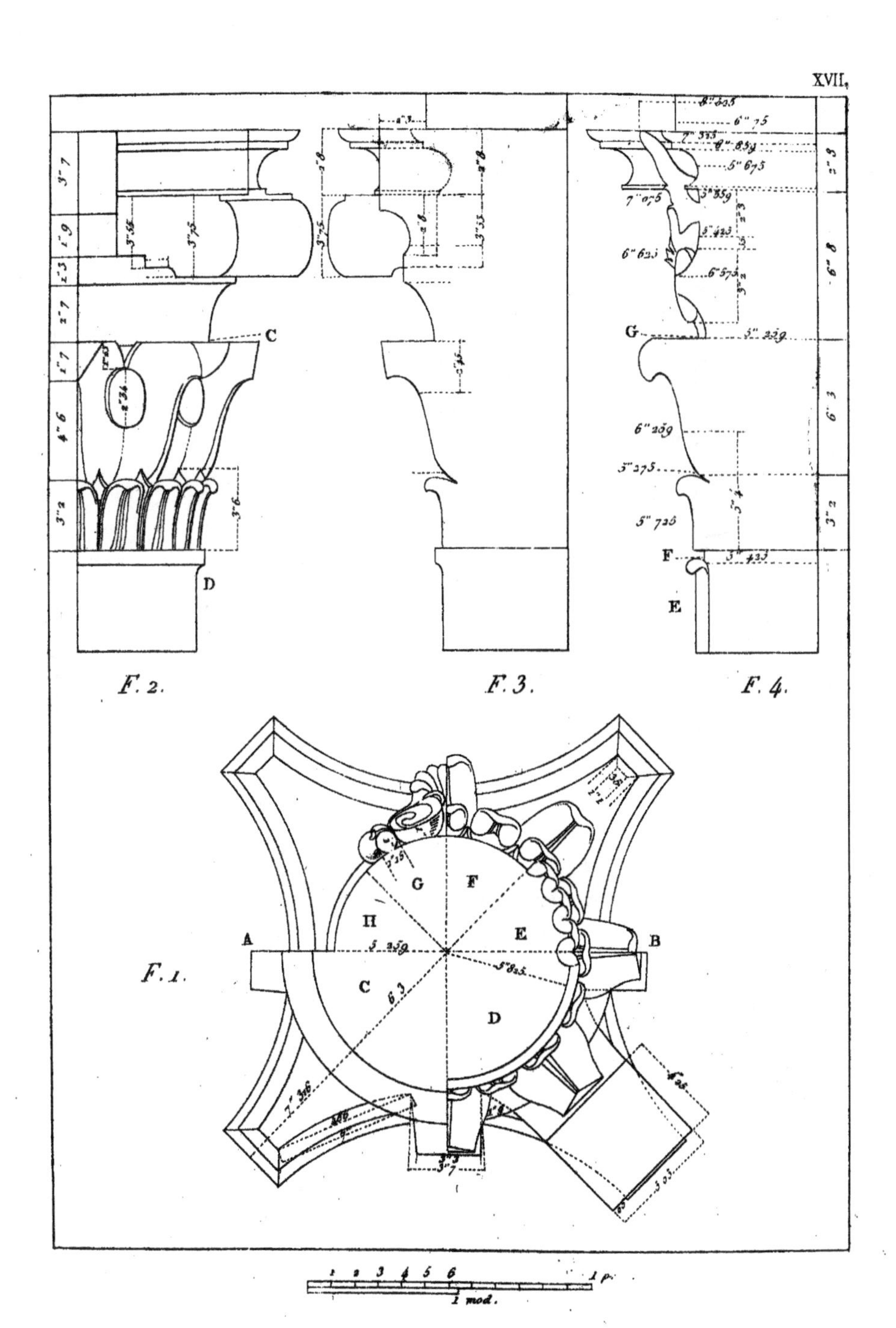
F. 2.
F. 3.
F. 4.
F. 1.
A
B
C
D
E
F
G
H
1 p.
1 mod.

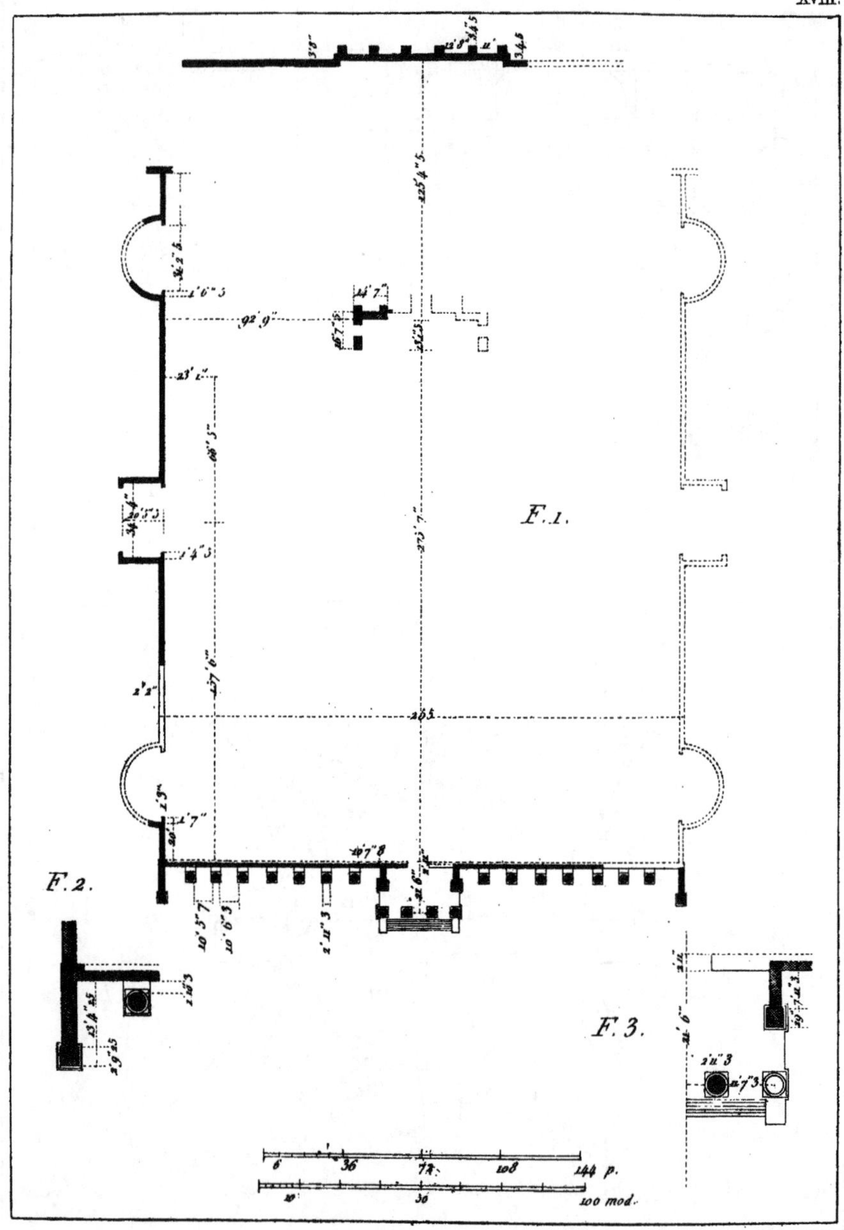
F. 1.
F. 2.
F. 3.
144 p.
100 mod.

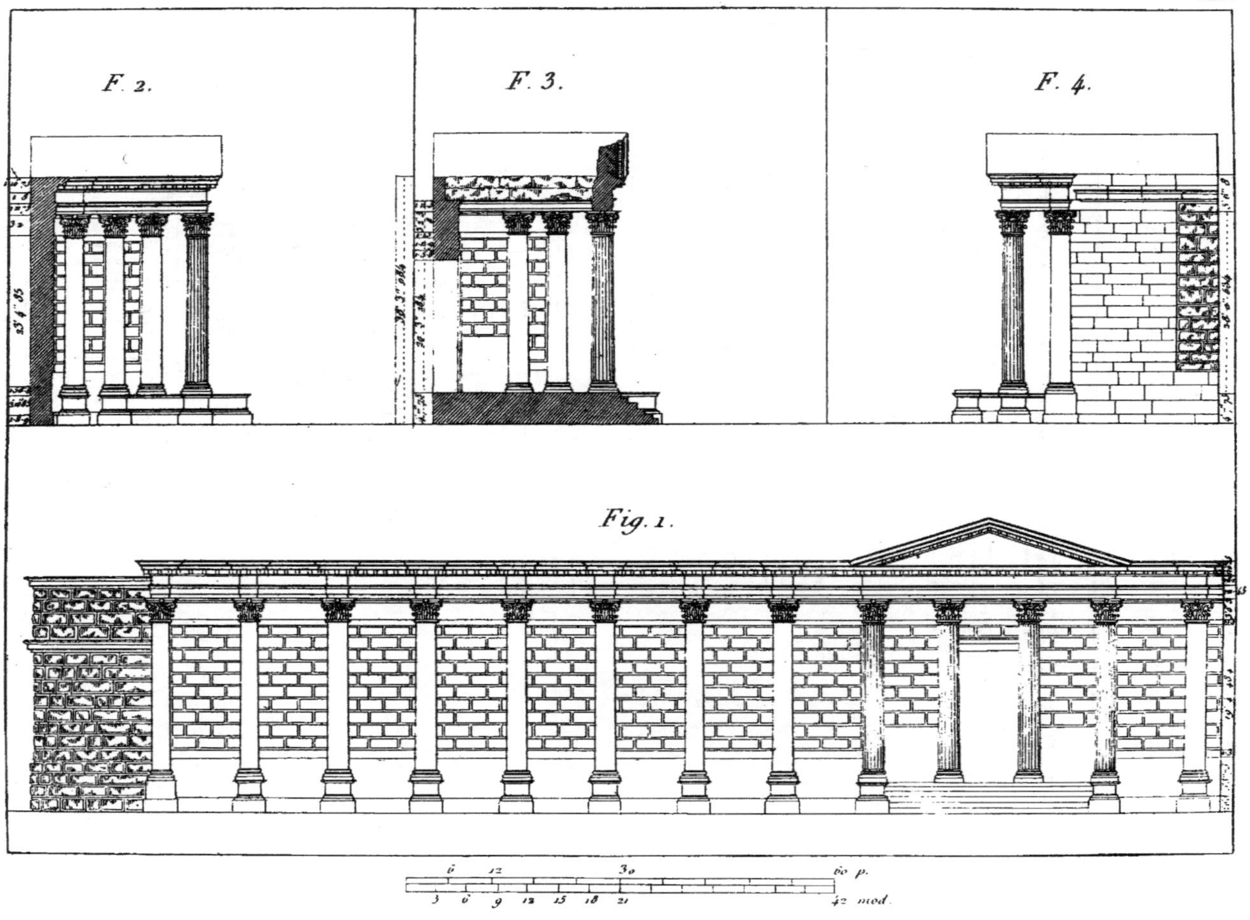

F. 2.

F. 3.

F. 4.

Fig. 1.

6 12 30 60 p.

3 6 9 12 15 18 21 42 mod.

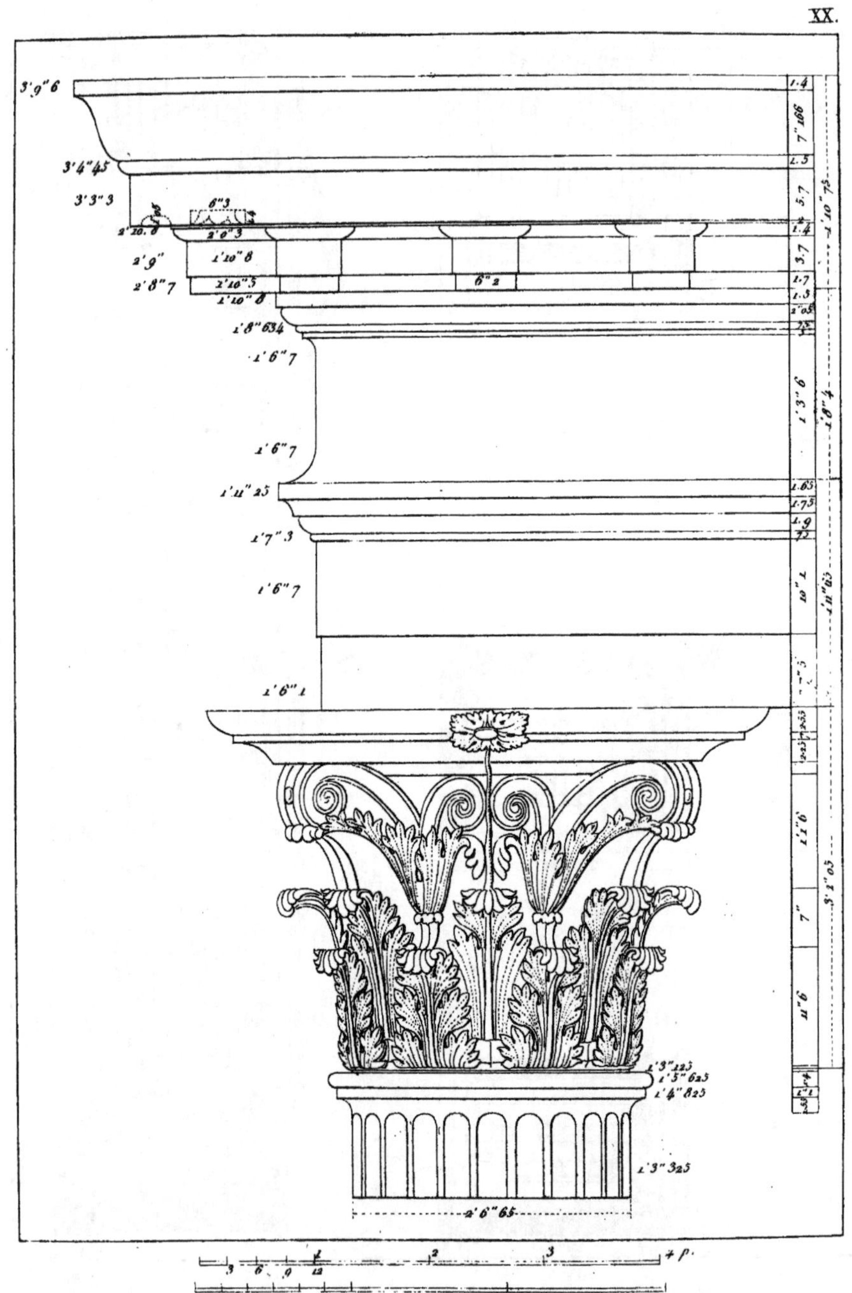
3'9"6
3'4"45
3'3"3
2'9"
2'8"7
1'6"7
1'6"7
1'11"25
1'7"3
1'6"7
1'6"1
1'3"325
2'6"65
4 p.
3 mod.

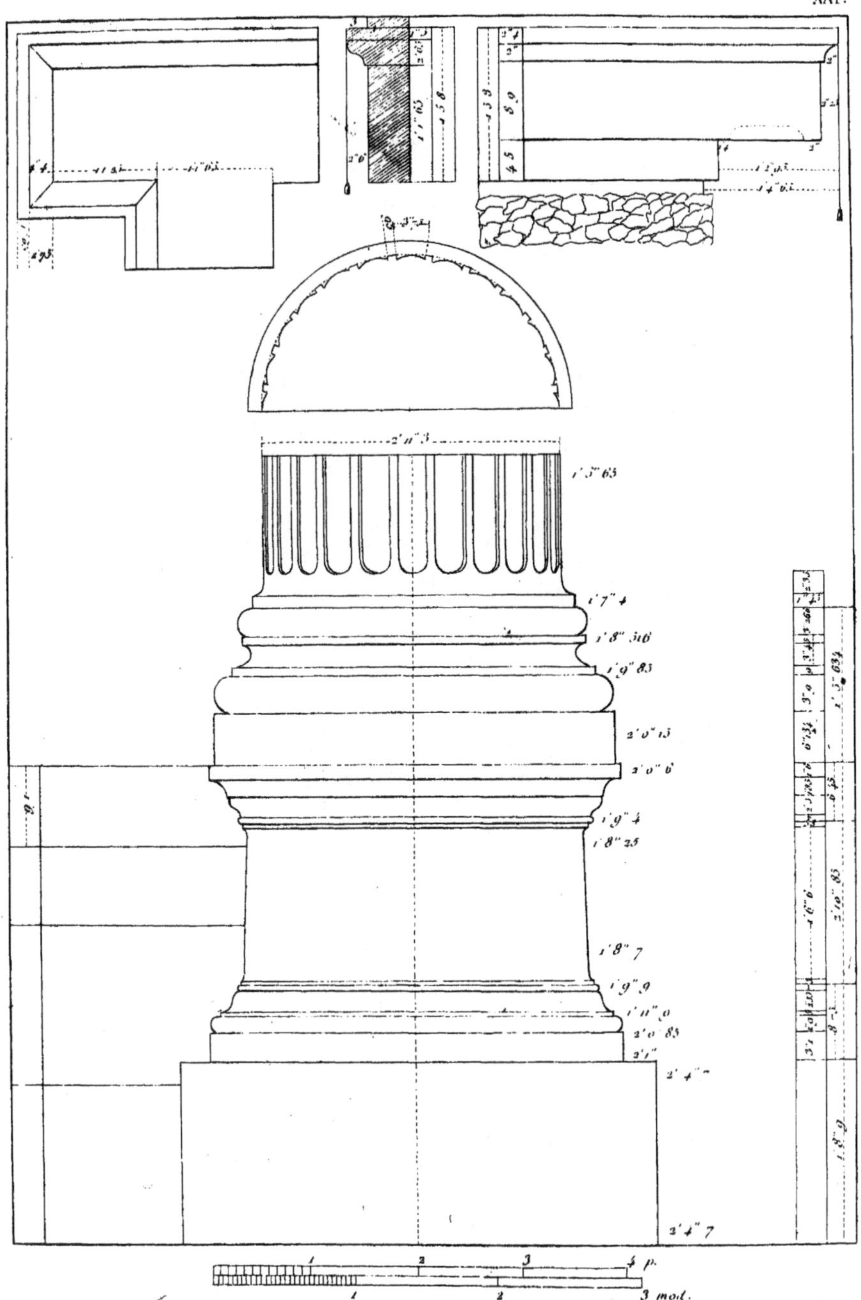

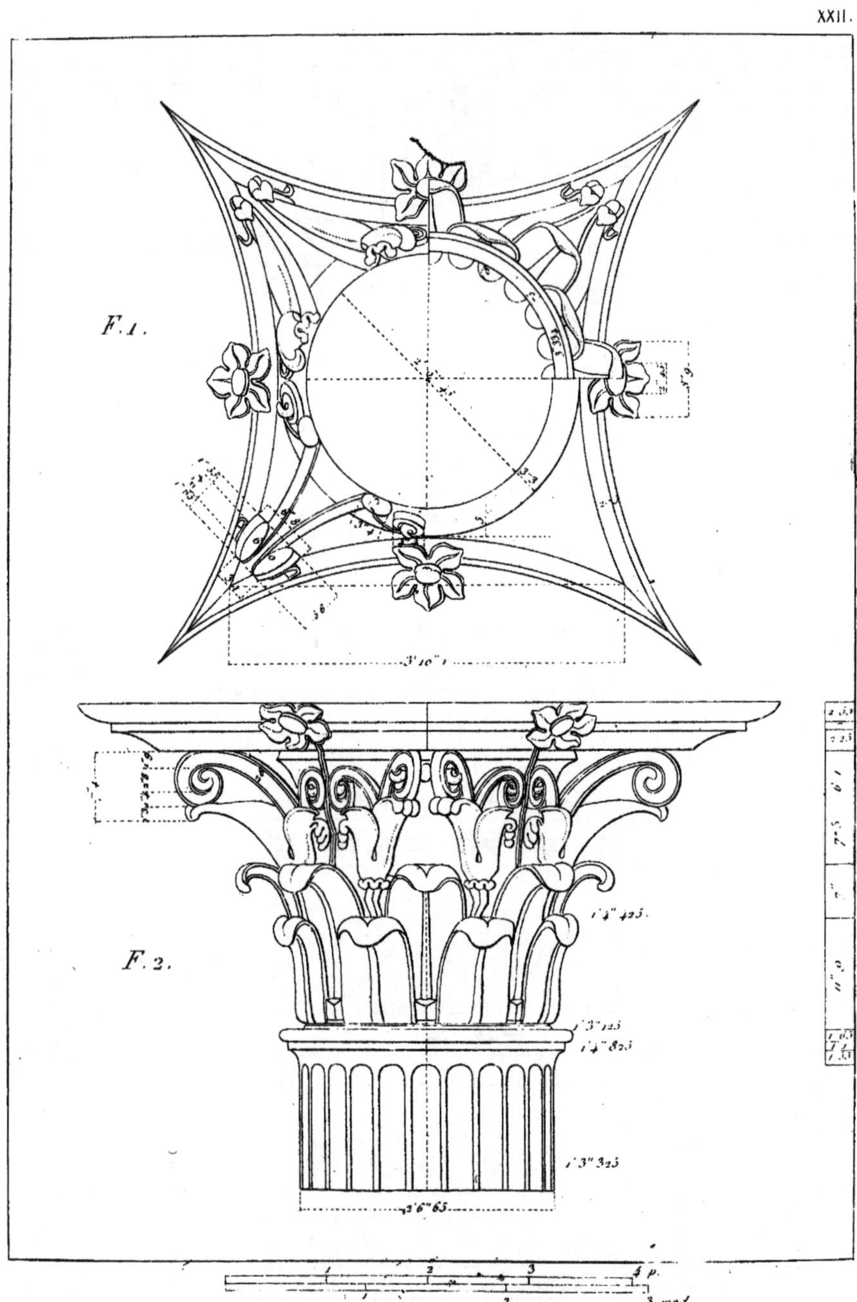
F. 1.
3' 10" 1
F. 2.
1' 4" 4 25
1' 3" 1 25
1' 4" 8 25
1' 3" 3 25
2' 6" 65
4 p.
3 mod.

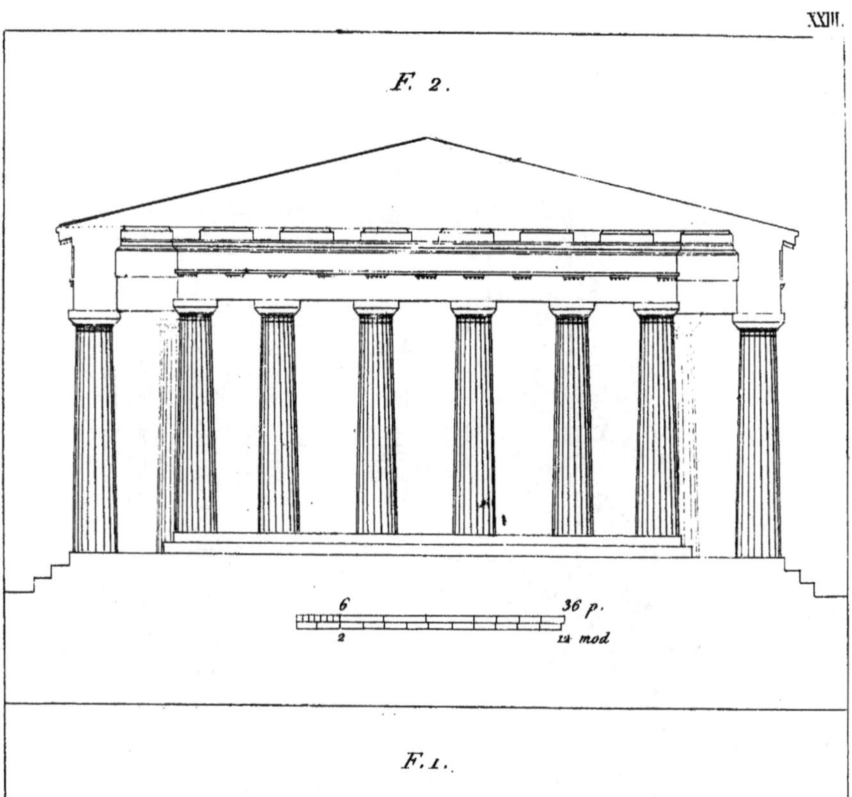
F. 2.
6
36 p.
2
12 mod

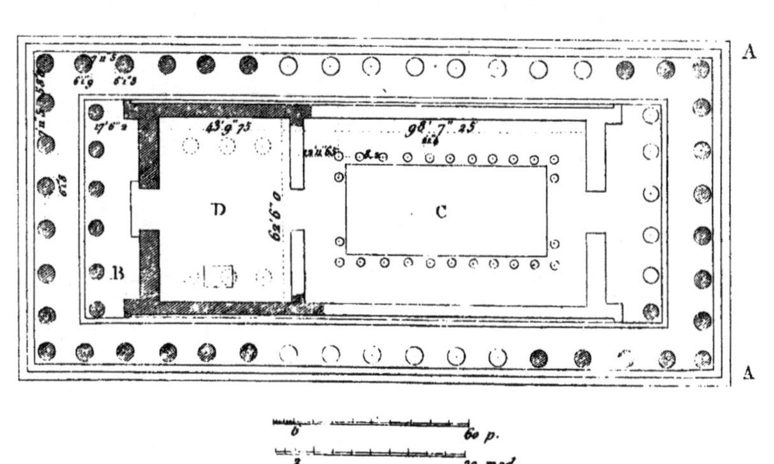
F. 1.
A
D
C
B
A
43' 9" 75
98' 7" 25
62' 6" 0
60 p.
20 mod.

XXIV.

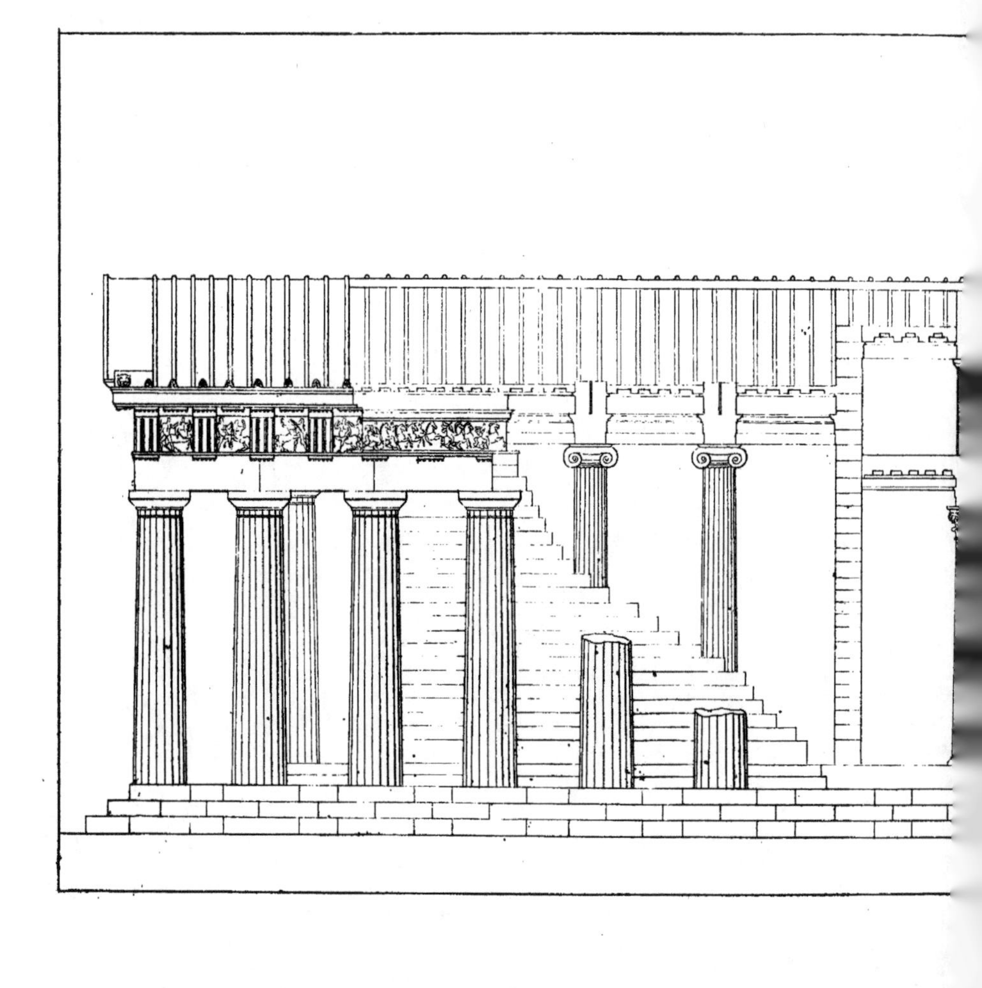

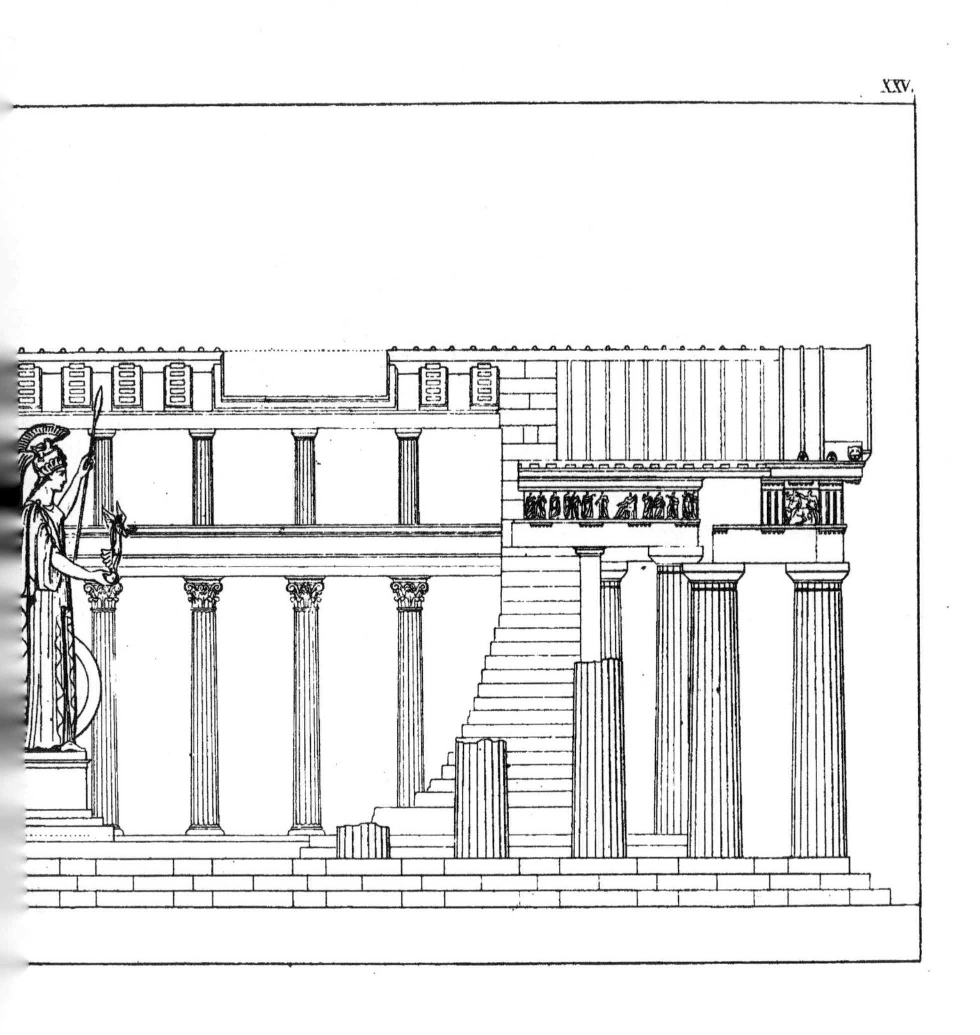

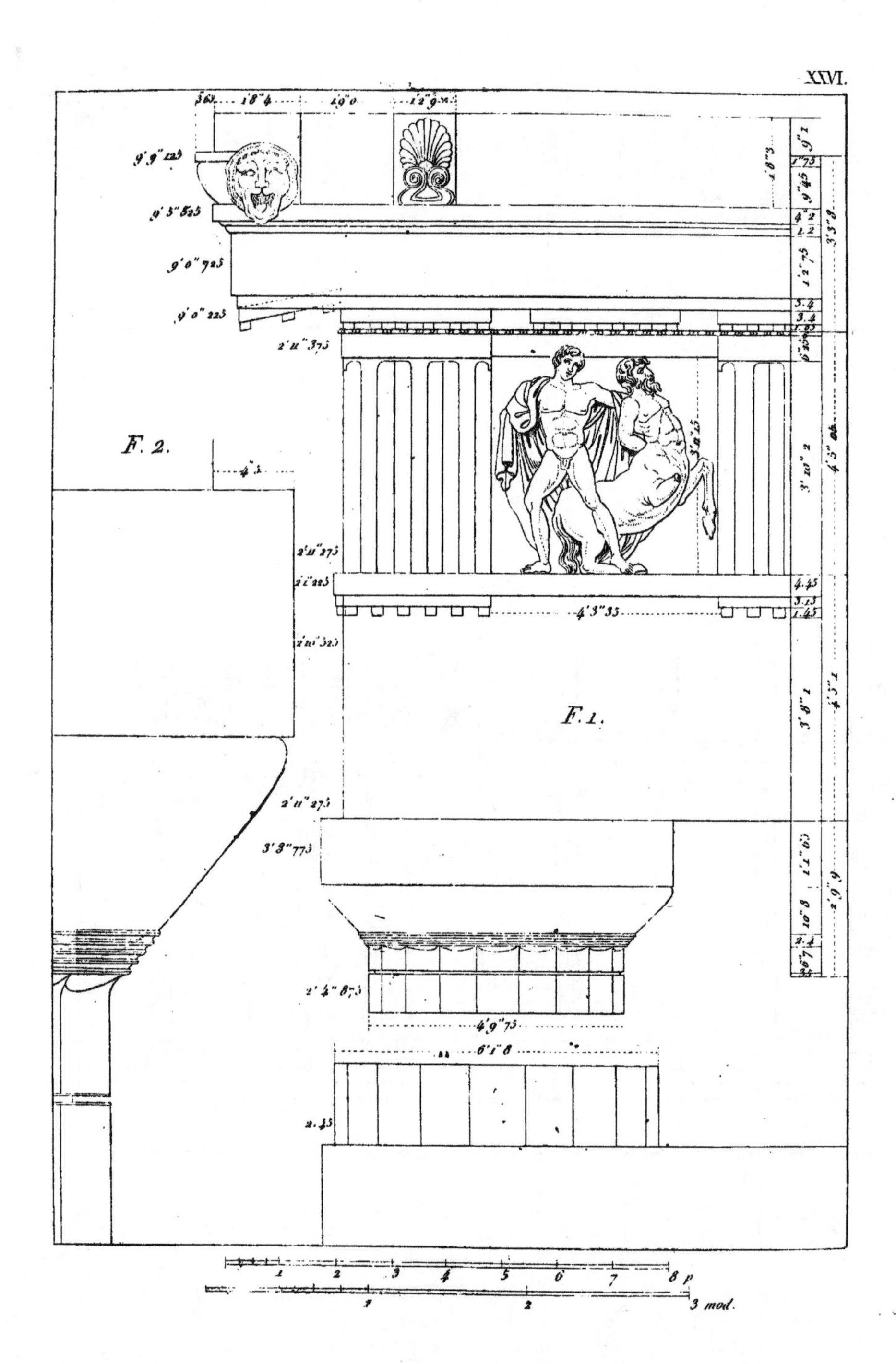
F. 1.
F. 2.
3 mod.
8 p

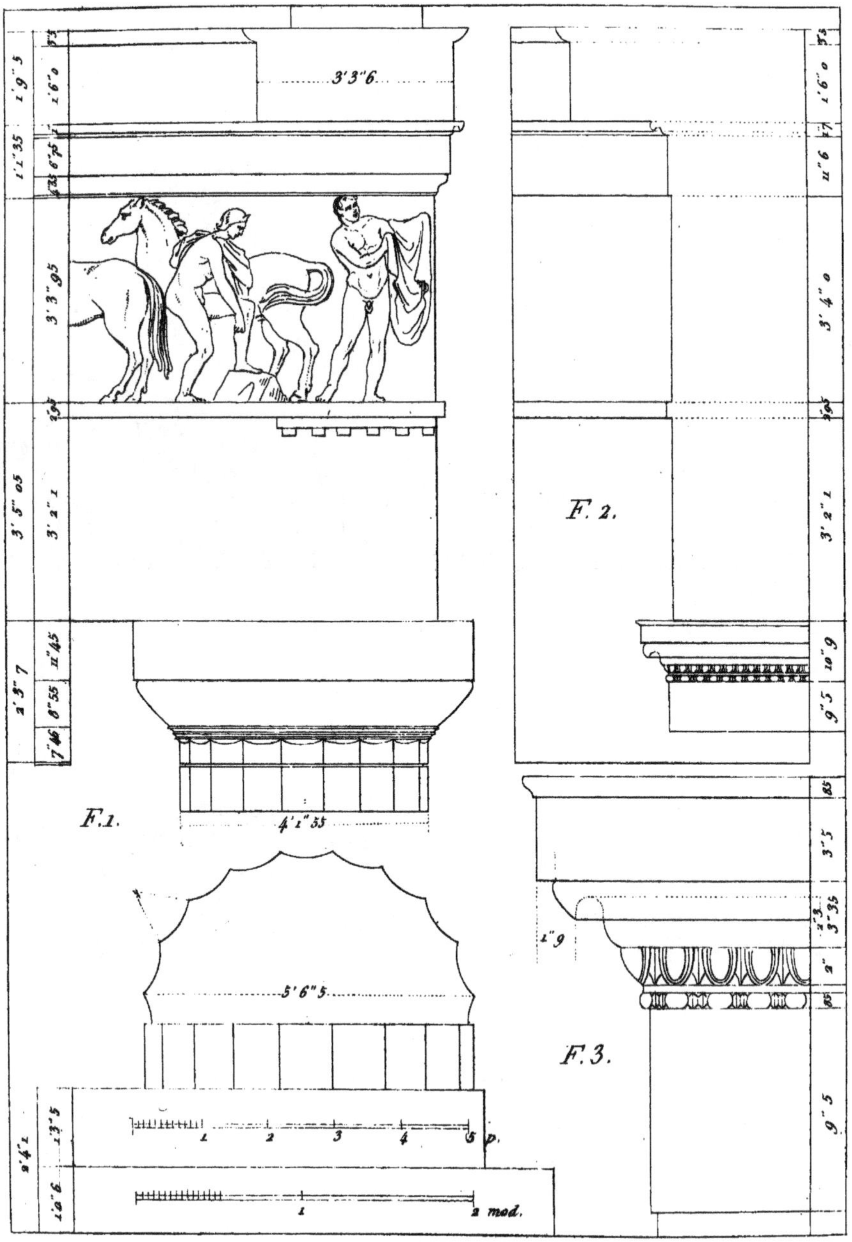

F. 1.

F. 2.

F. 3.

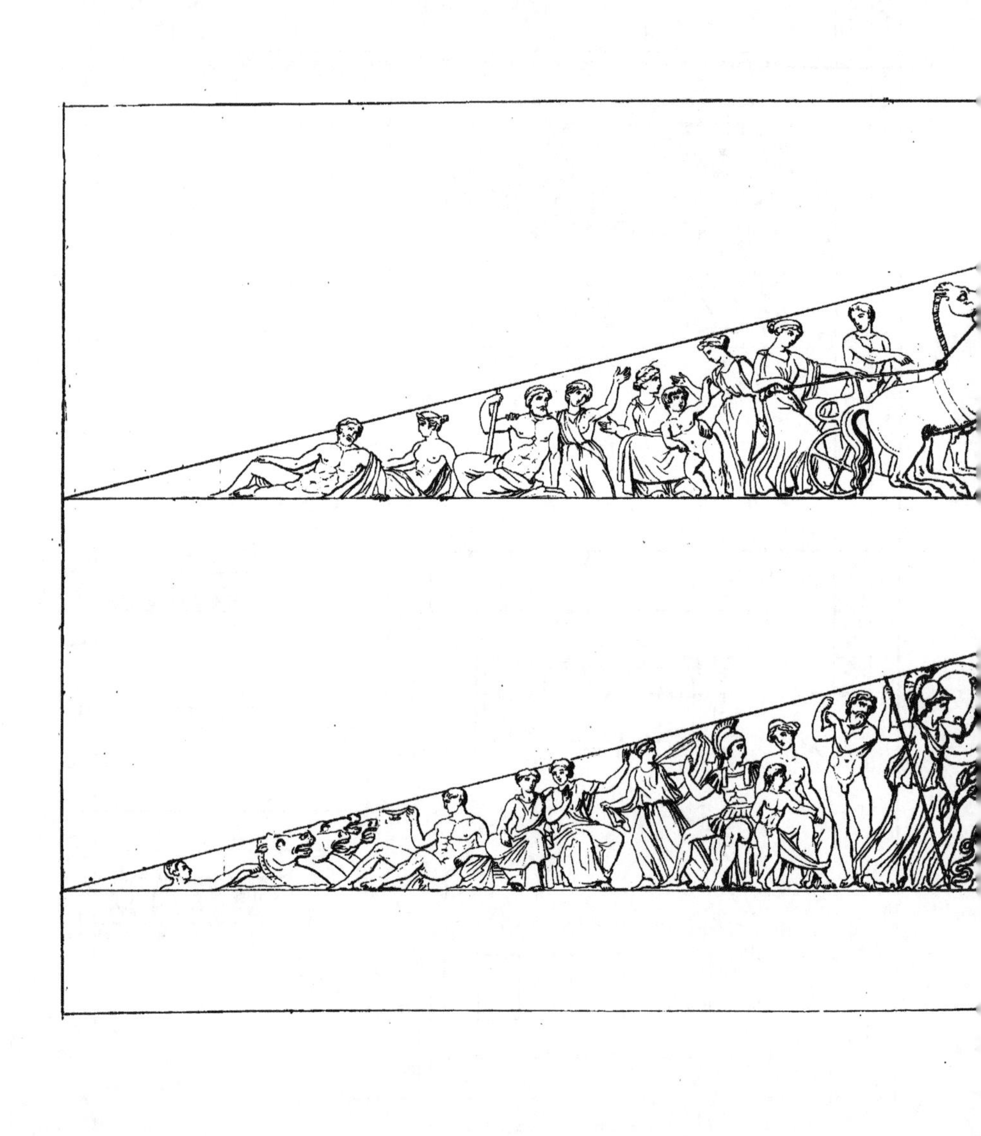

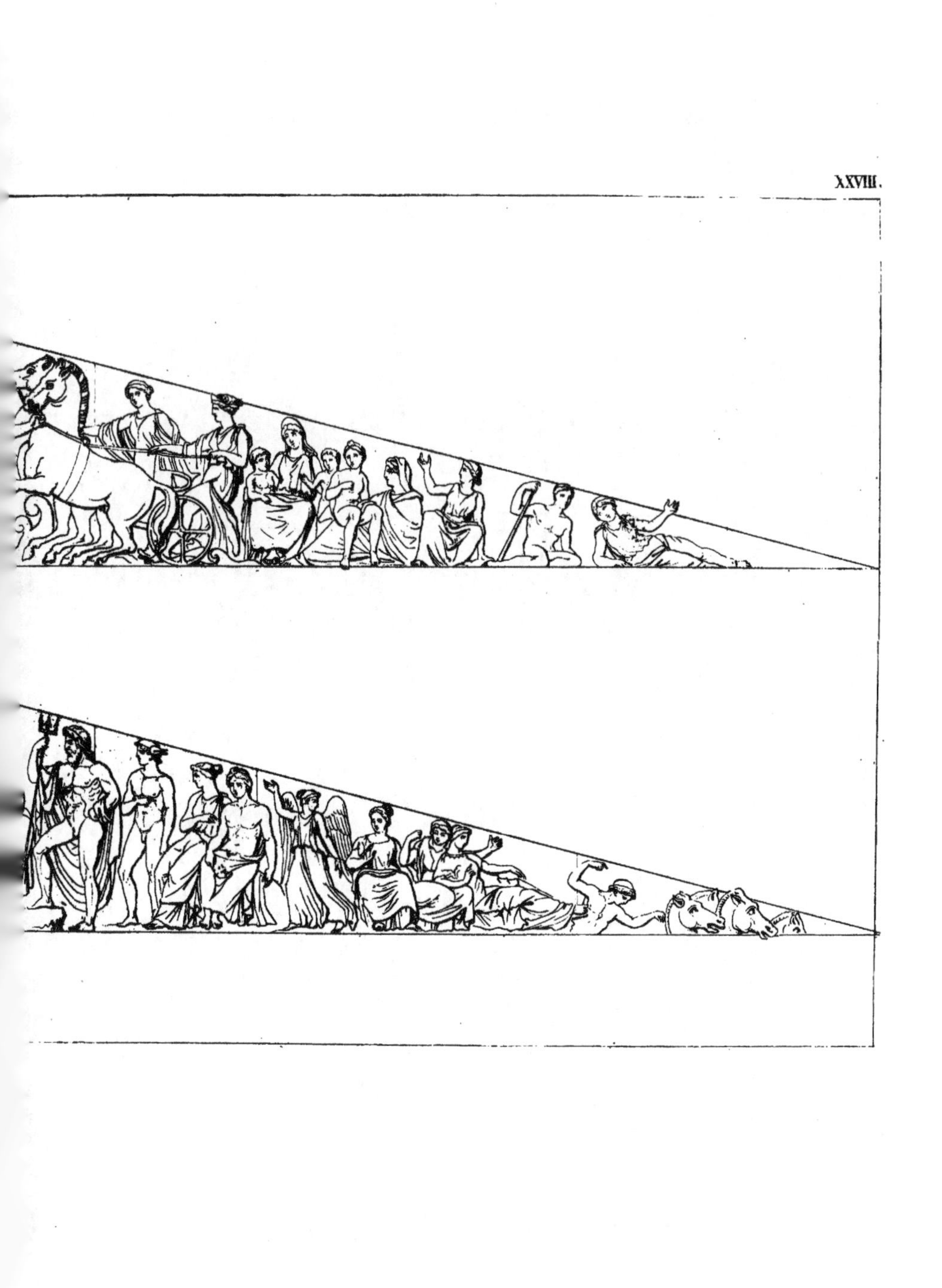

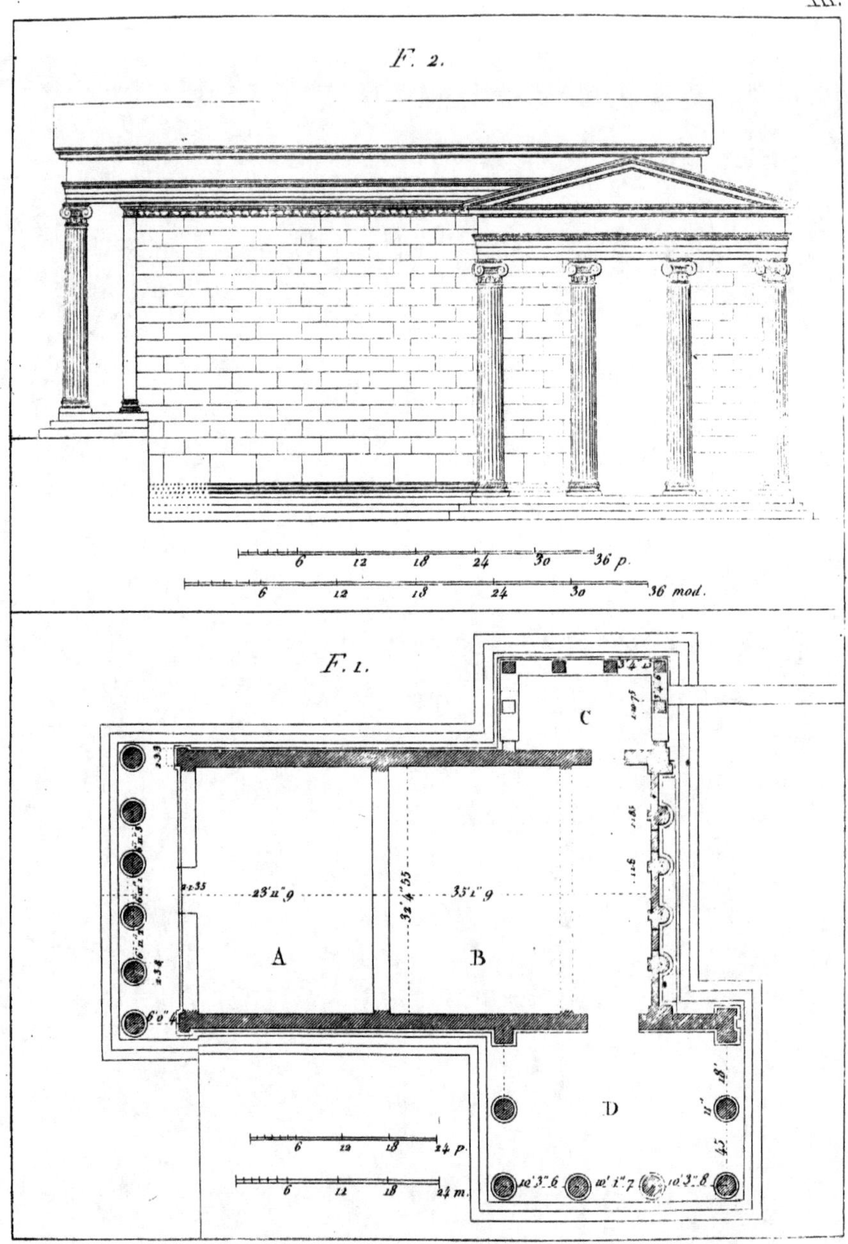
F. 2.
6 12 18 24 30 36 p.
6 12 18 24 30 36 mod.
F. 1.
A
B
C
D
23' 11" 9
35' 1" 9
6 12 18 24 p.
6 12 18 24 m.

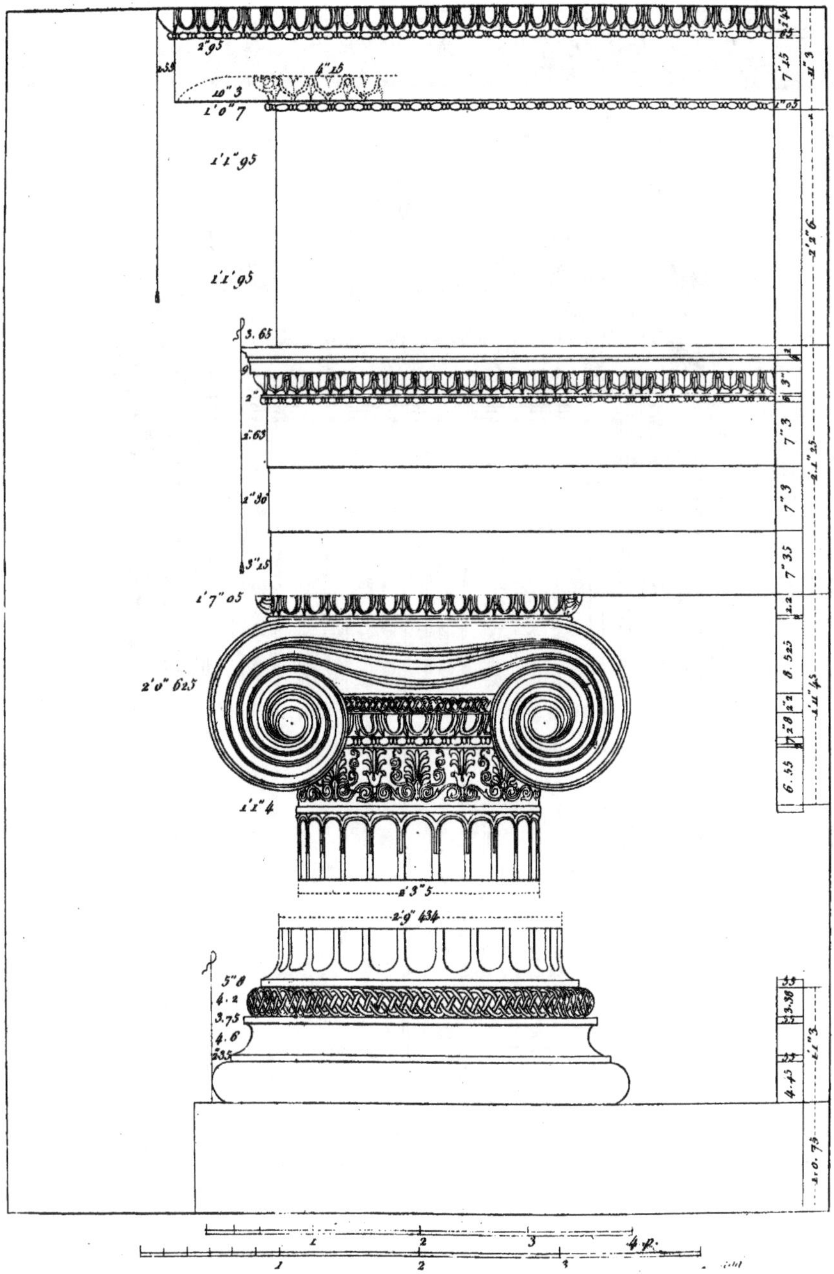

XXXI.

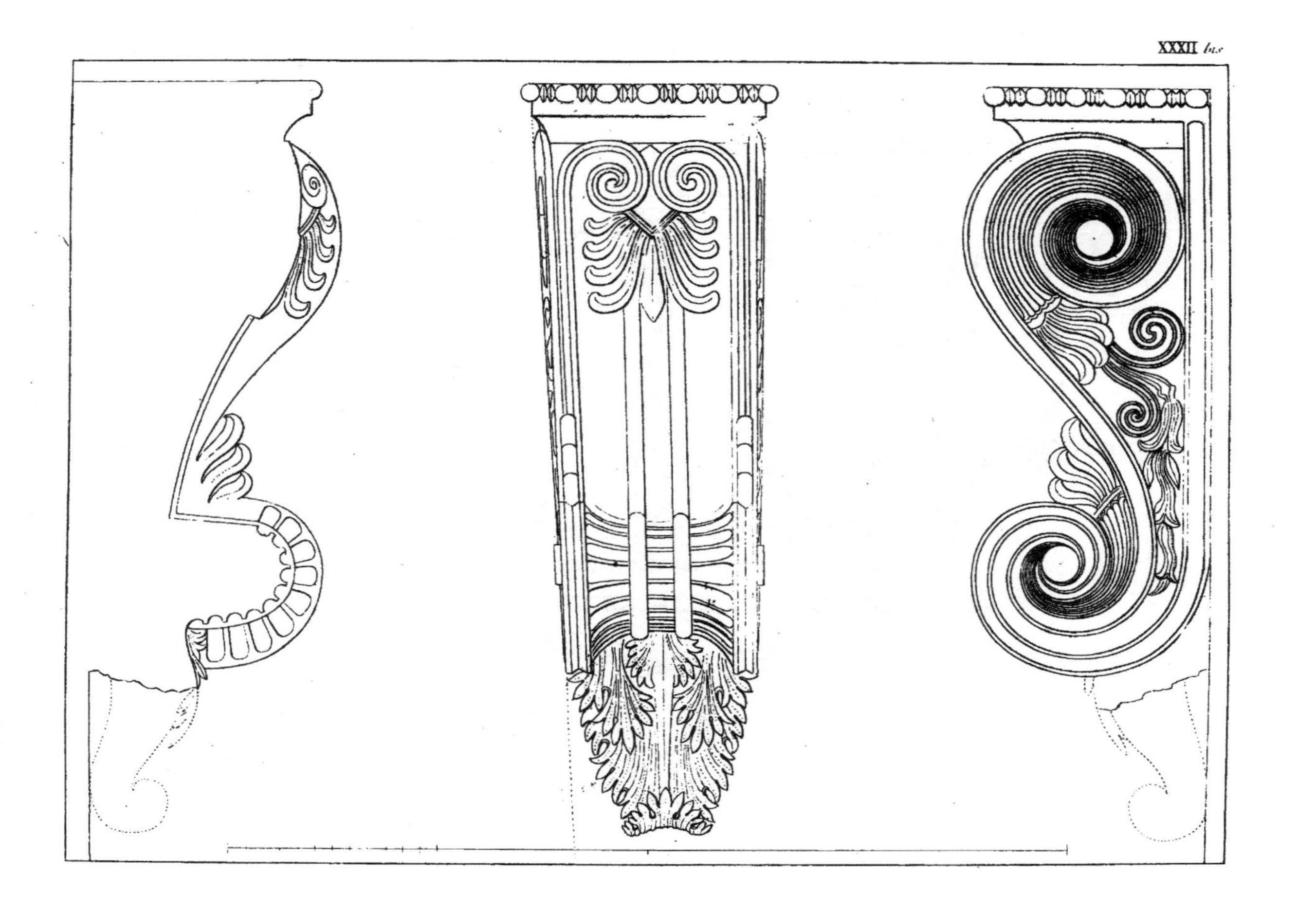

XXXIII.

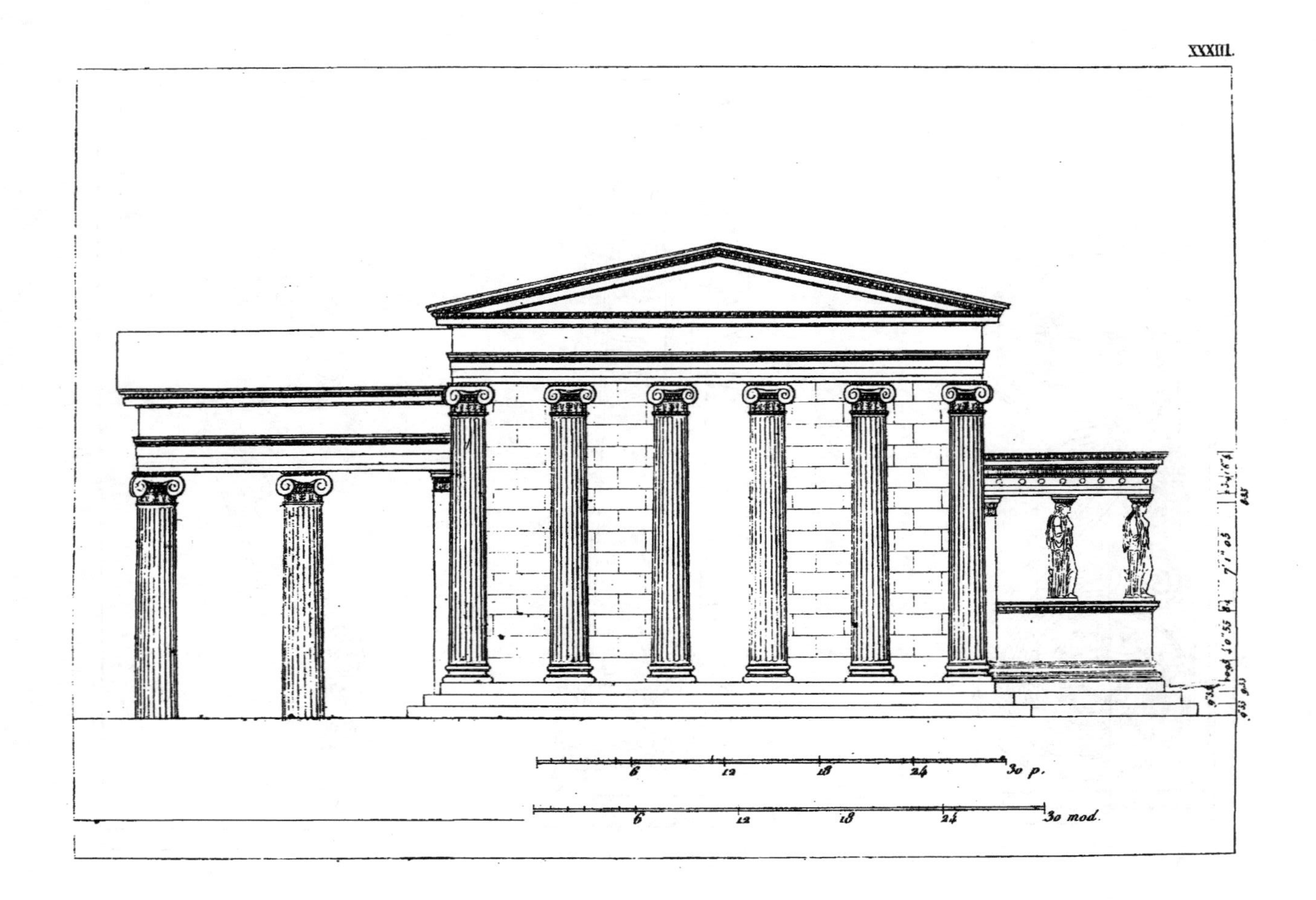

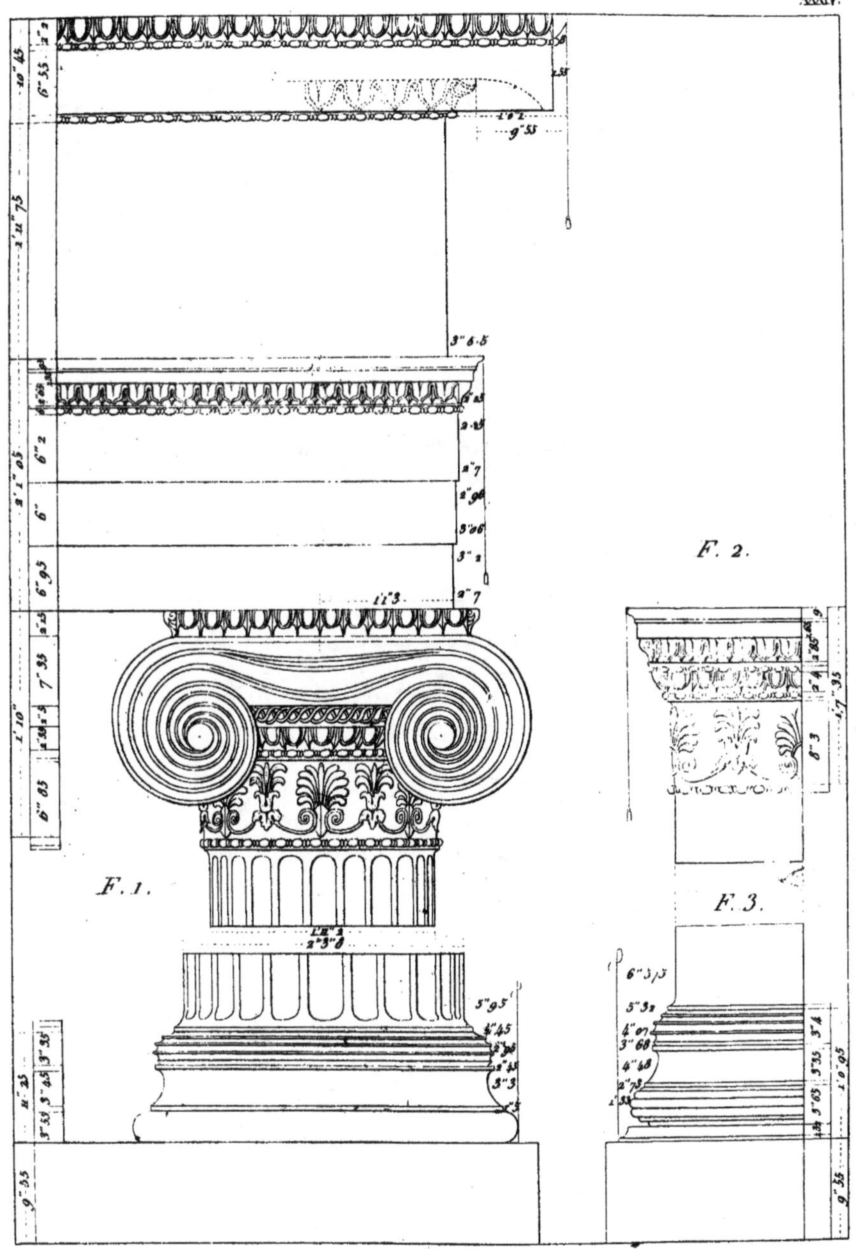
F. 1.
F. 2.
F. 3.

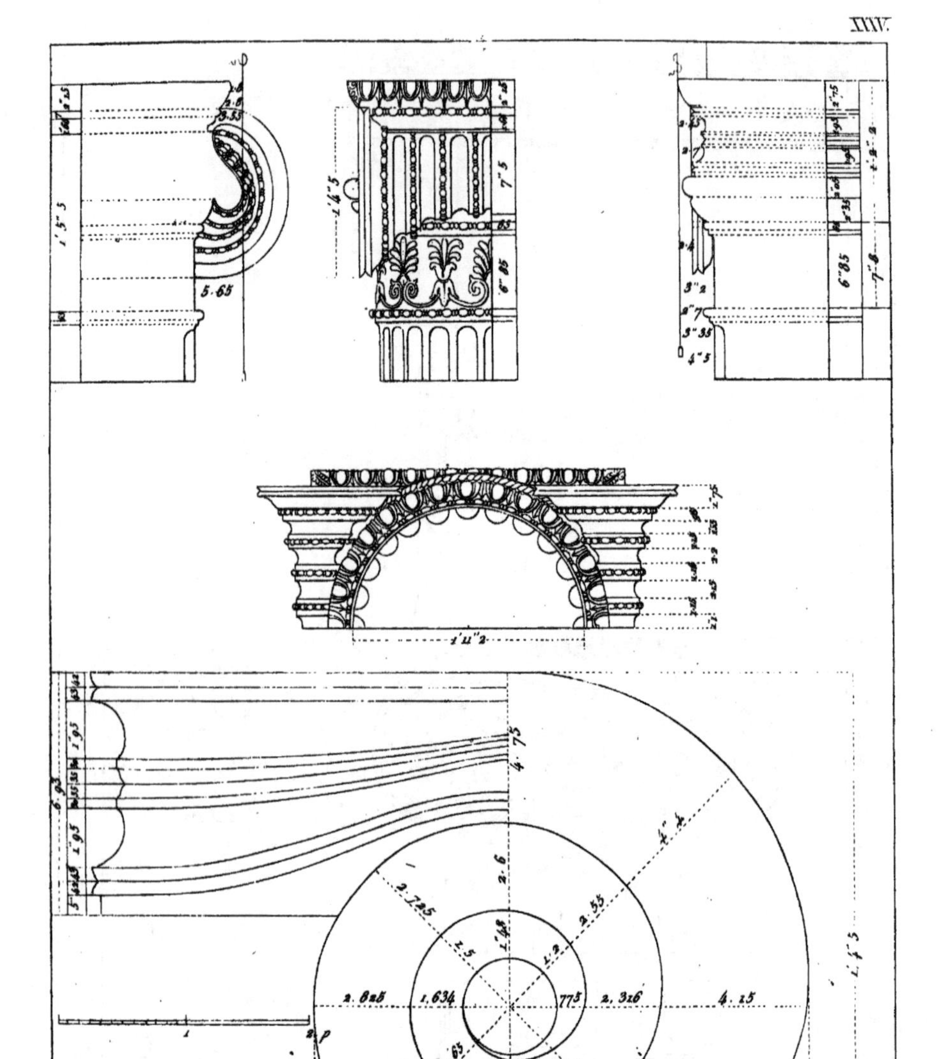

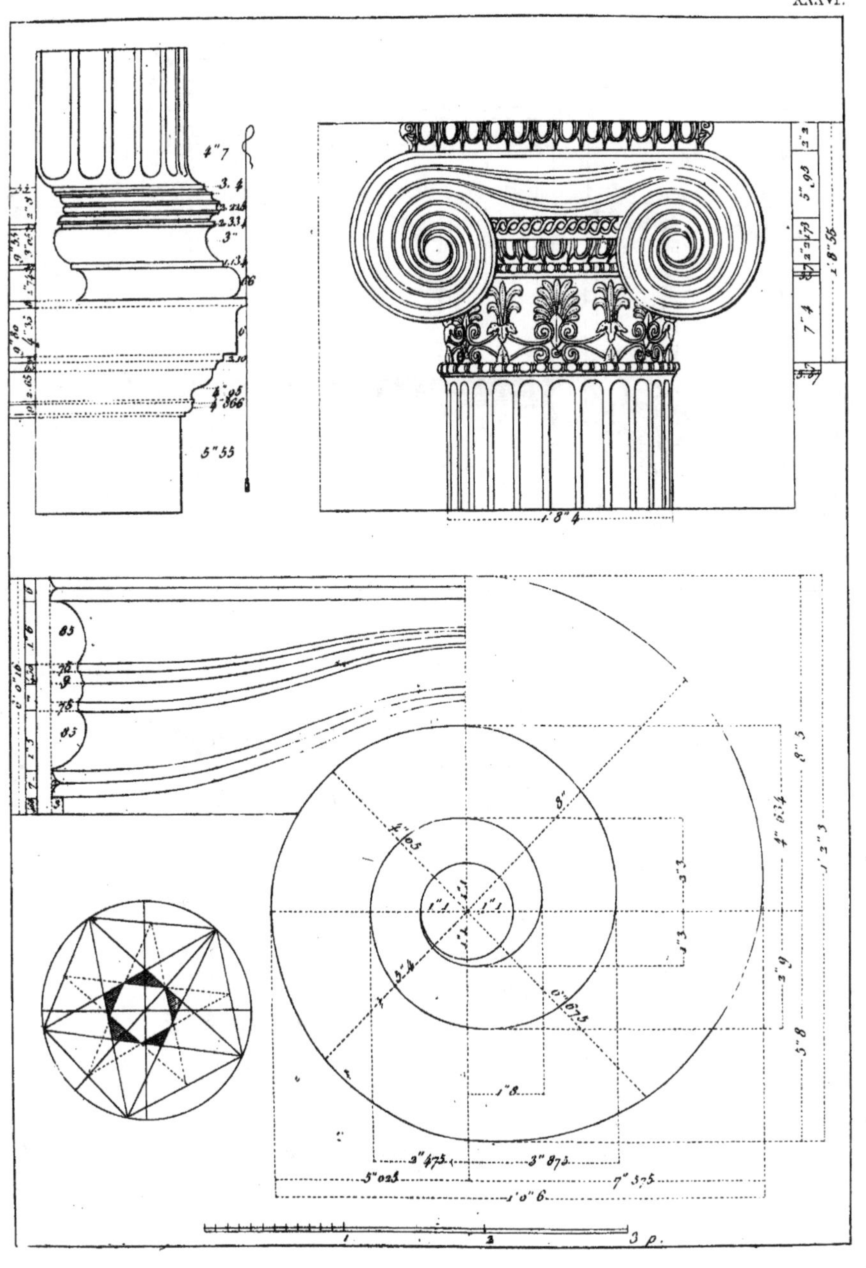

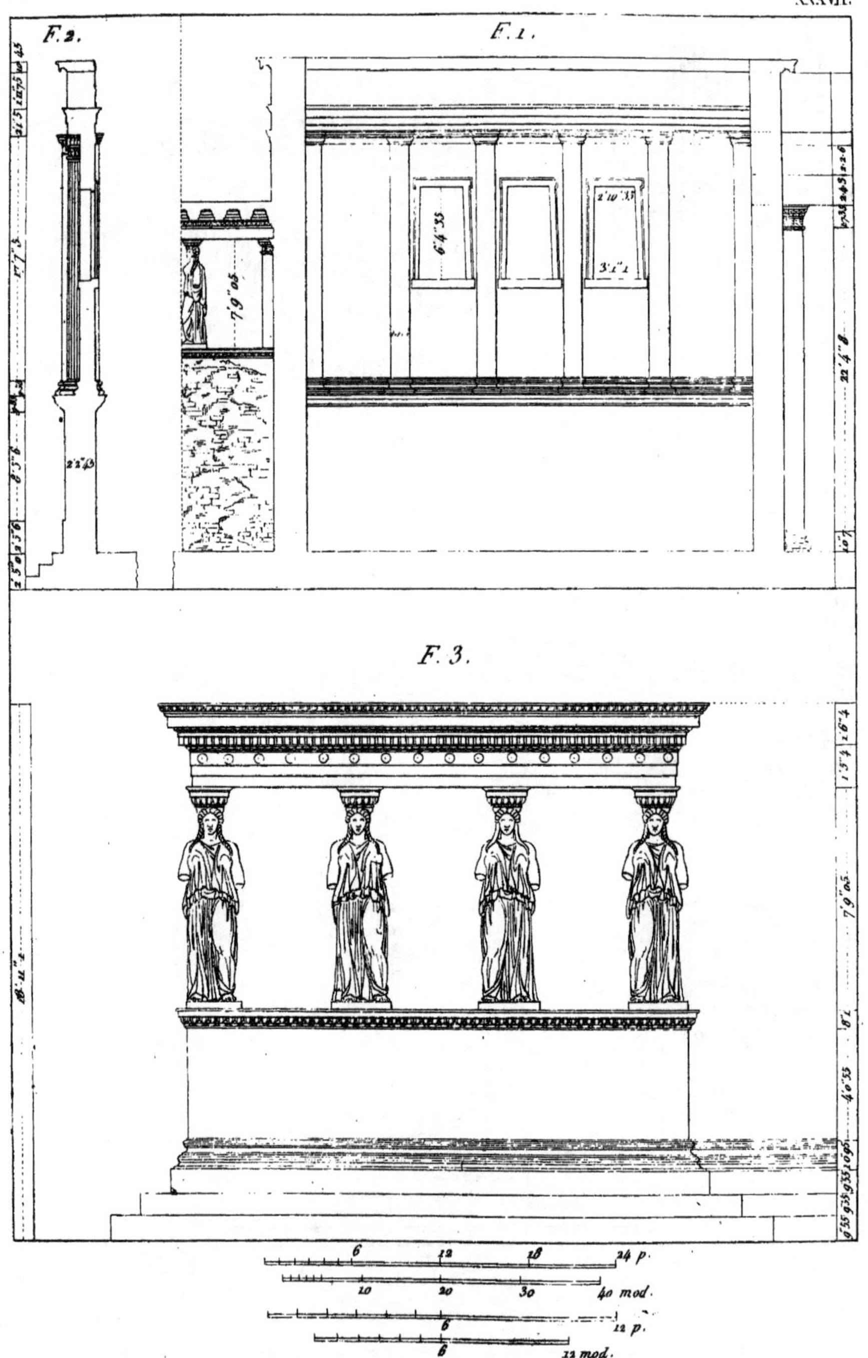
F. 2.
F. 1.
F. 3.
6 12 18 24 p.
10 20 30 40 mod.
6 12 p.
6 12 mod.

XXXVIII

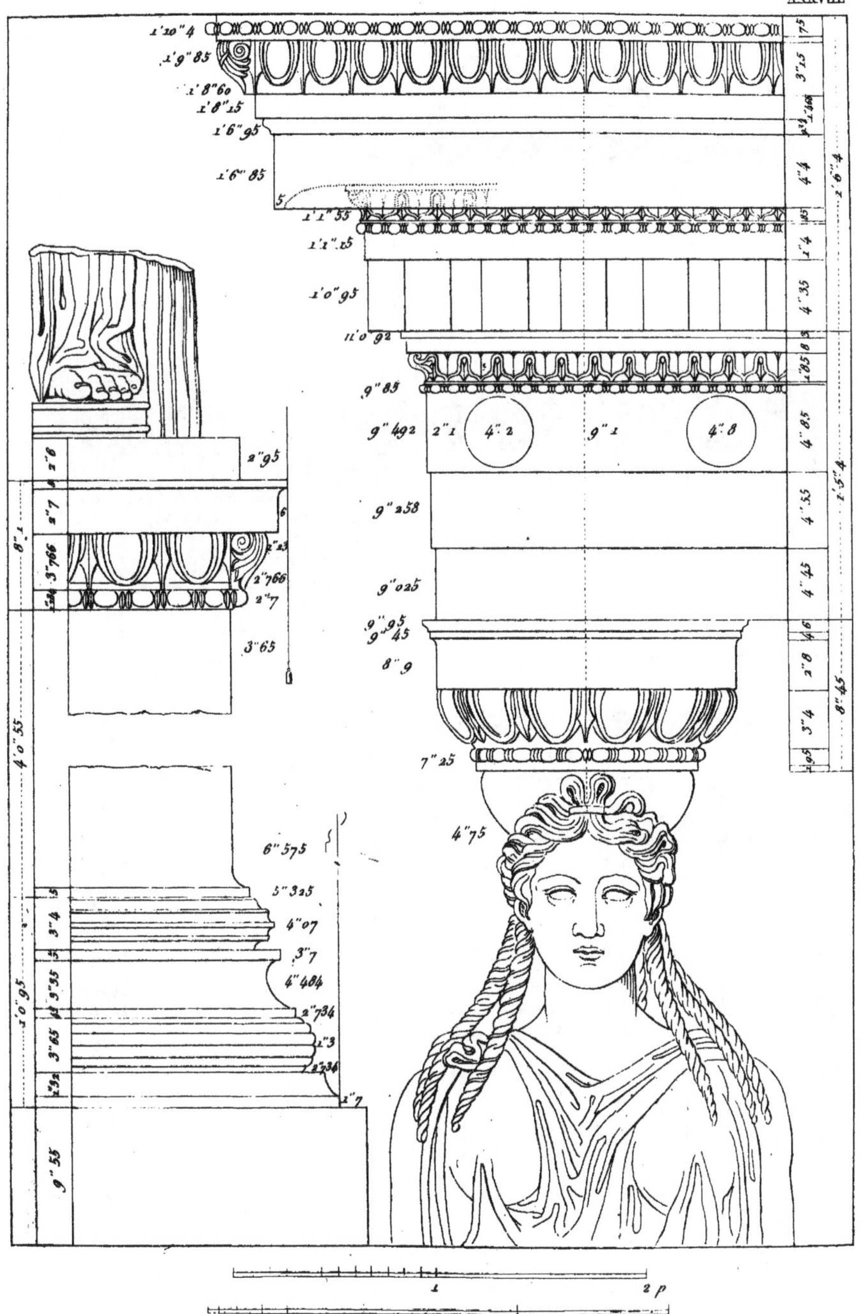

1'10"4
1'9"85
1'8"60
1'8"15
1'6"95
1'6"85
5
1'1"55
1'1".15
1'0"95
11'0"92
9"85
9"492
2"1
4".2
9".1
4".8
9"258
9"025
9"95
9"45
8"9
7"25
4"75
6"575
5"325
4"07
3"7
4"484
2"734
1"3
2"234
2"7
2"95
6
2"766
2"7
3"65
75
3"15
1'6"4
4"4
1"4
4"35
4"85
1'5"4
4"55
4"45
2"8
3"4
8"45
8"1
2"8
2"7
3"766
4'0"55
3"4
3"35
3"65
1'0"95
9"55
1
2 p
1
2
3 mod

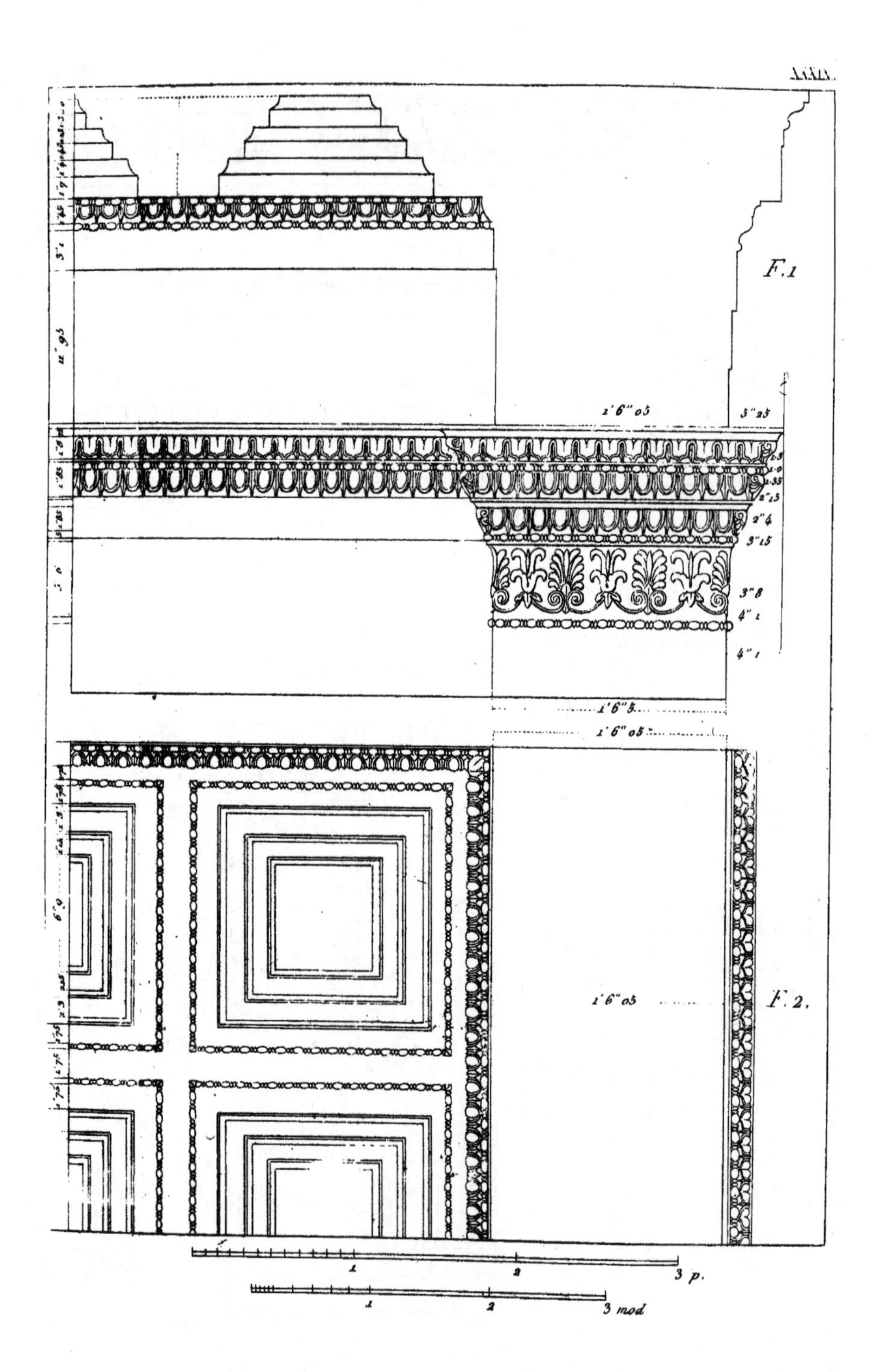
F. 1
F. 2.
1' 6" o5
3" 25
1' 6" 5
1 2 3 p.
1 2 3 mod

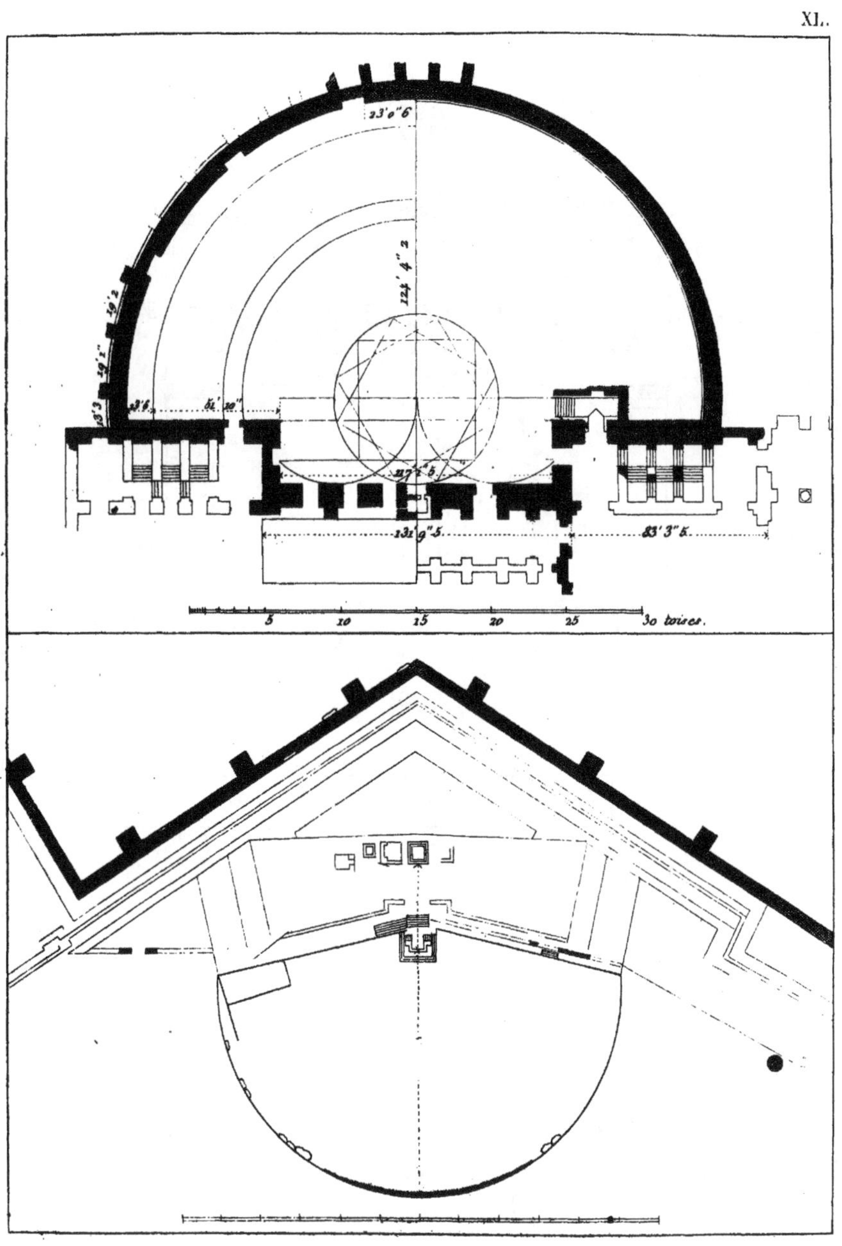
23' 0" 6
124' 4" 2
19' 2
19' 1"
13' 3
23' 6
61' 10"
127' 2" 5
131' 9" 5.
83' 3" 5.
5
10
15
20
25
30 toises.

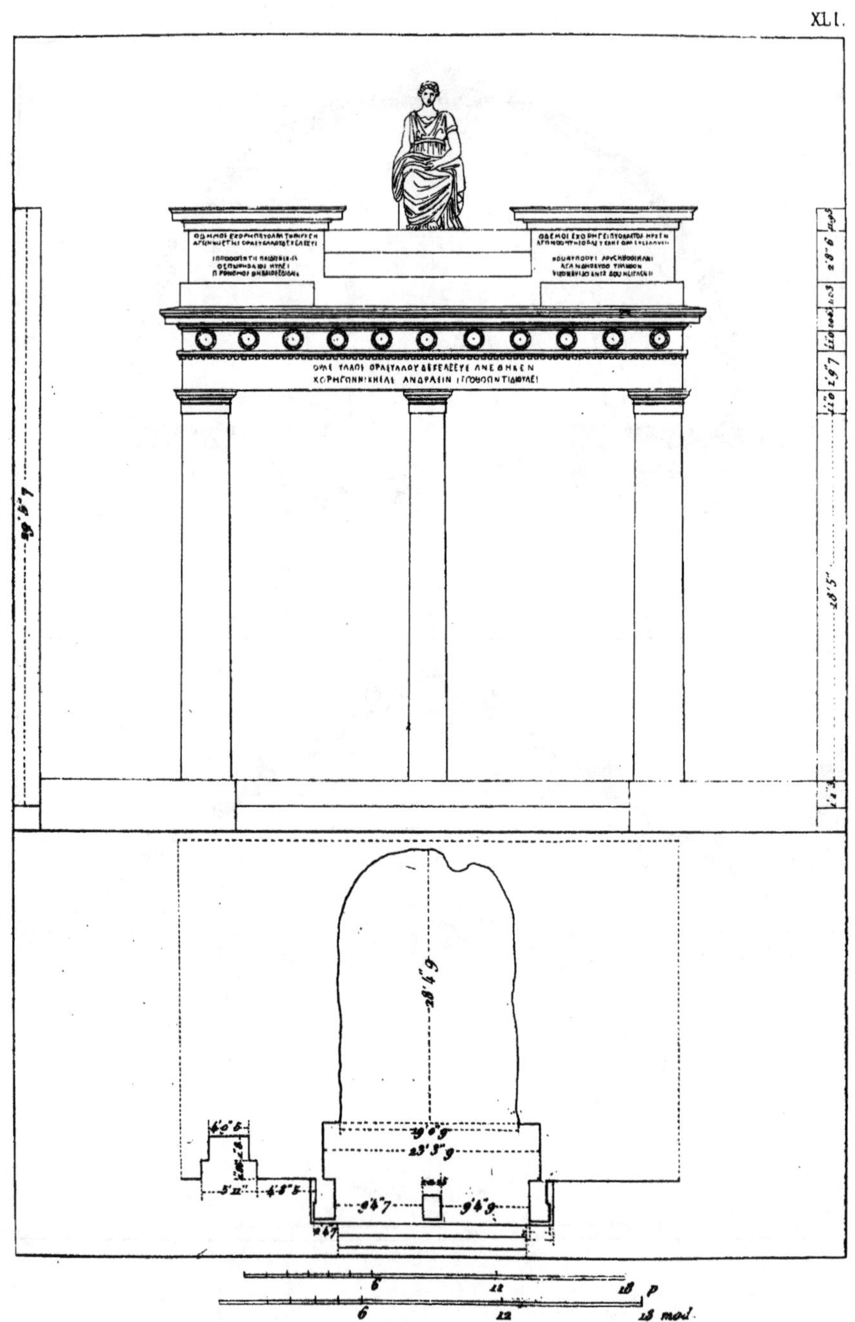

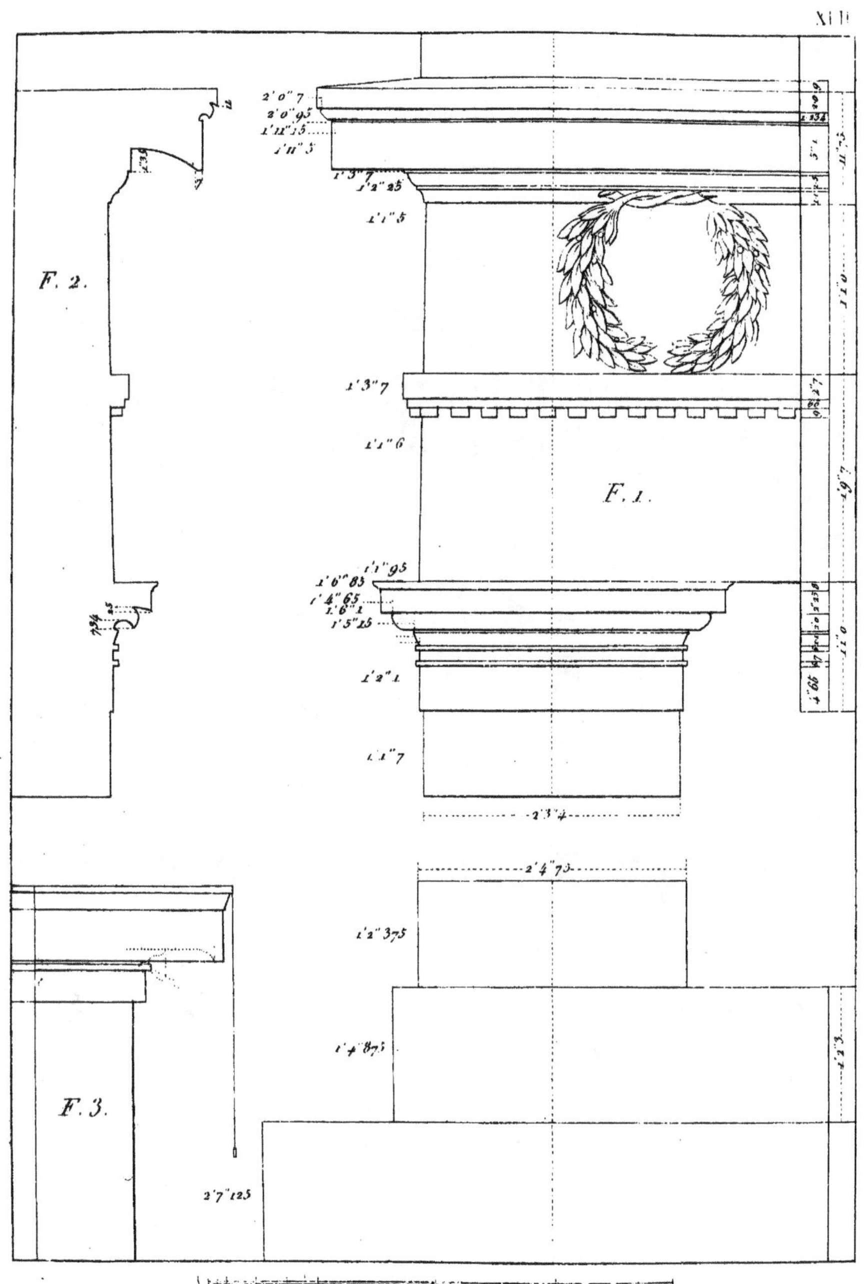
F. 1.
F. 2.
F. 3.
2'0"7
2'0"95
1'11"15
1'11"5
1'3"7
1'2"25
1'1"5
1'3"7
1'1"6
1'1"95
1'6"83
1'4"65
1'6"1
1'5"15
1'2"1
1'1"7
2'3"4
2'4"73
1'2"375
1'4"875
2'7"125
1 3 4 p.
1 2 3 mod.

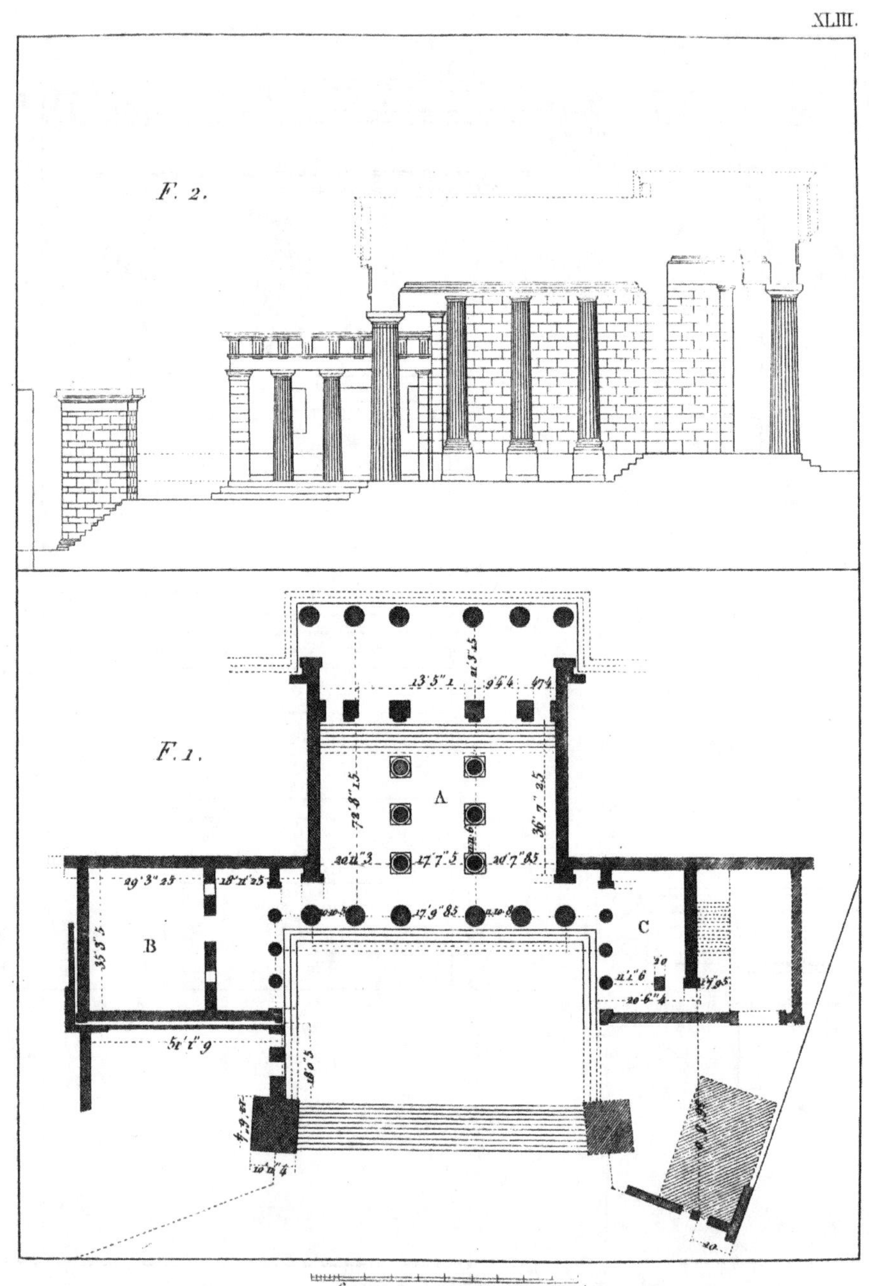
F. 2.
F. 1.
A
B
C
6
00 p.
6
12
18
24 mod.

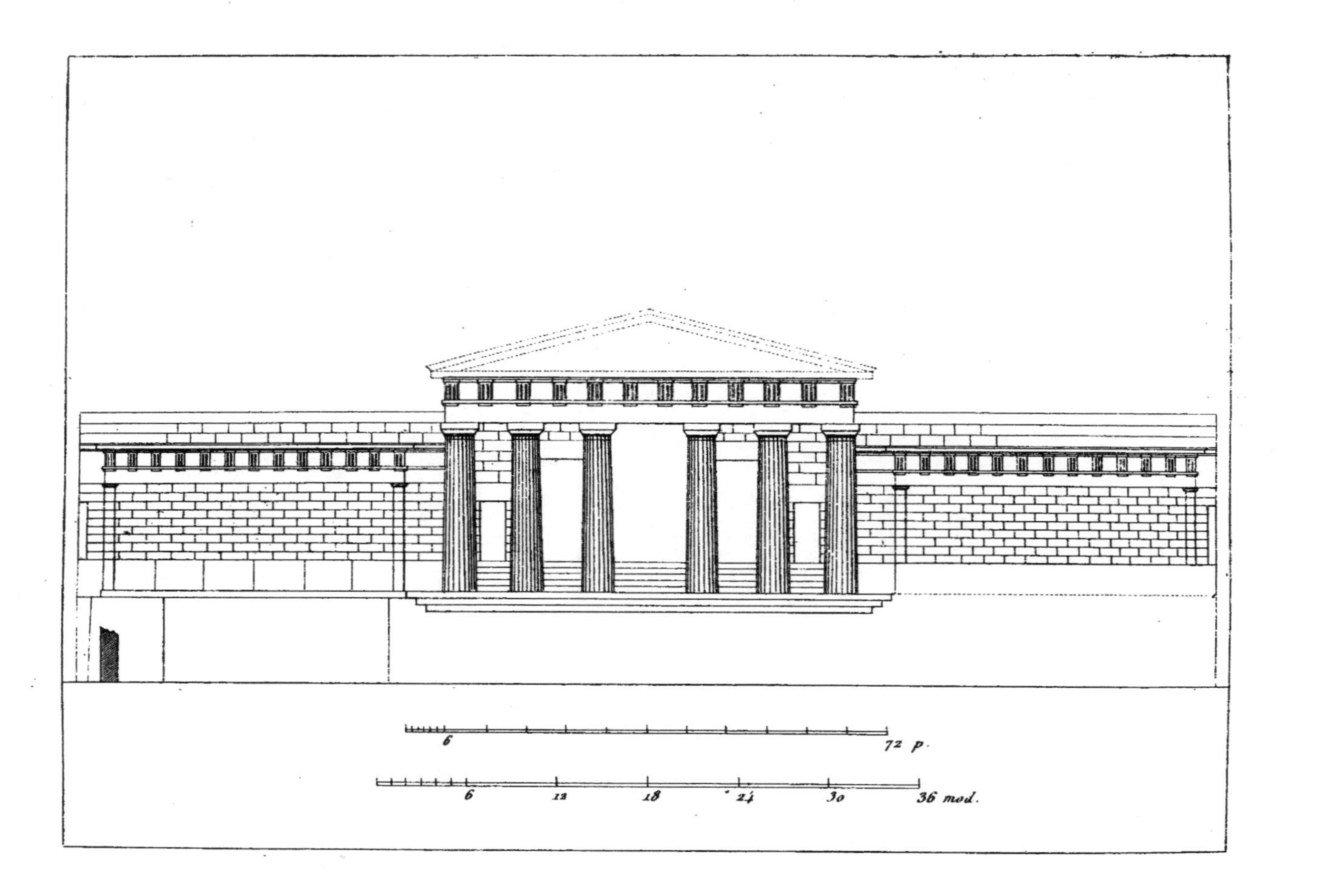
6
72 p.
6
12
18
24
30
36 mod.

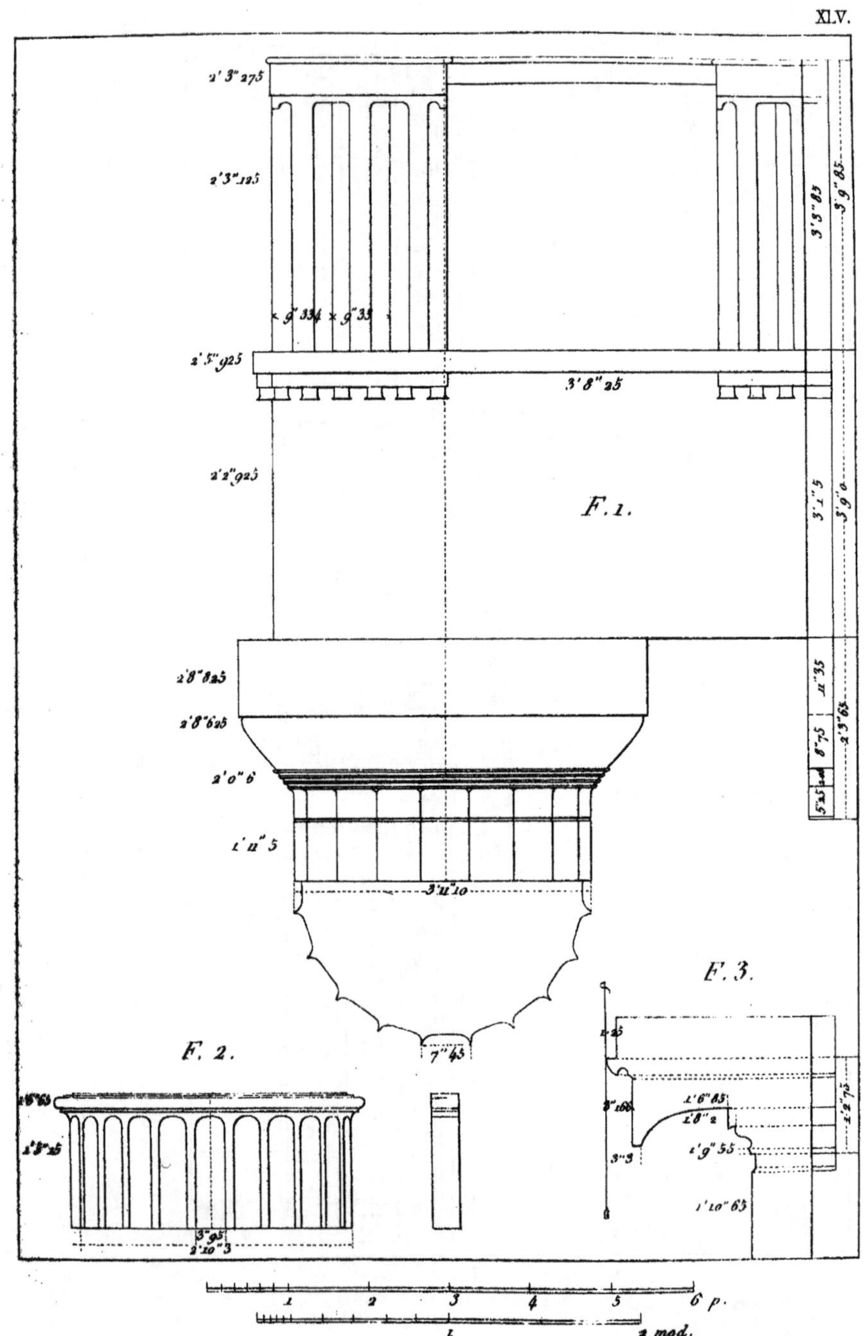
F. 1.
F. 2.
F. 3.
6 p.
2 mod.

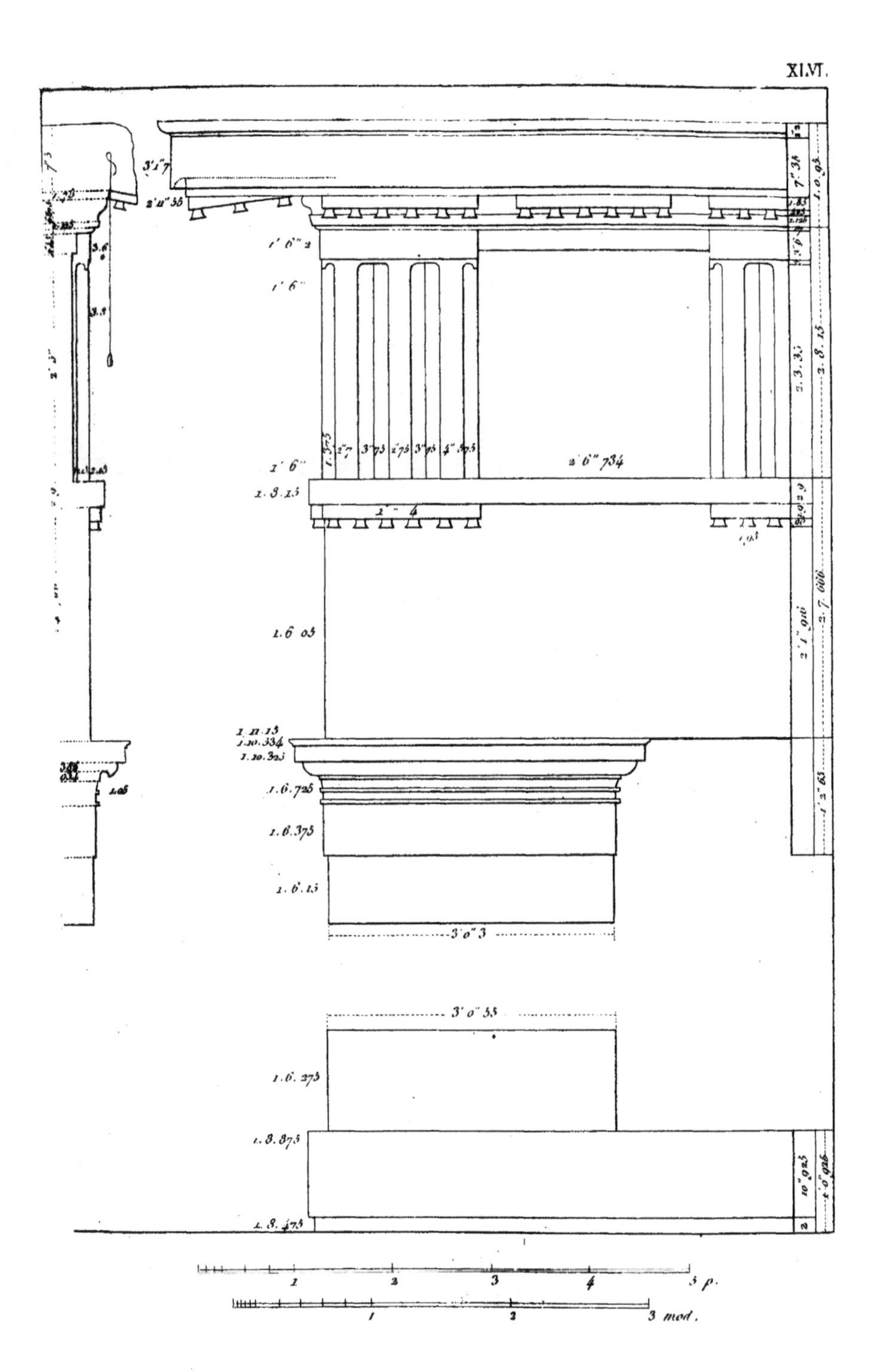

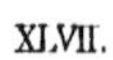
XLVII.

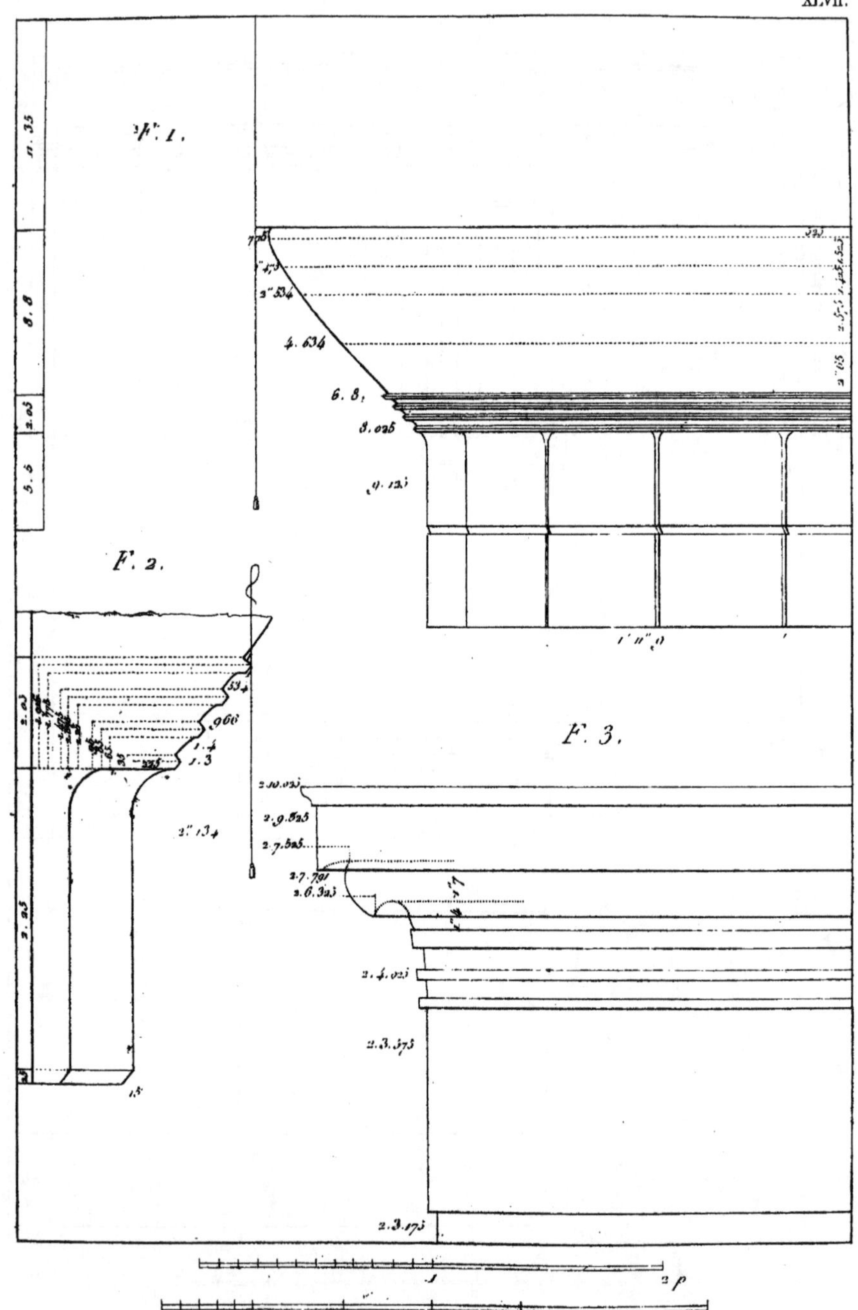
F. 1.
F. 2.
F. 3.
4. 634
6. 8
8. 026
5. 5
2. 23
8. 8
11. 35
2.10.023
2.9.823
2.7.523
2.7.791
2.6.323
2.4.023
2.3.373
2.3.173
1
2 p
5
10
15
20
30 mod.

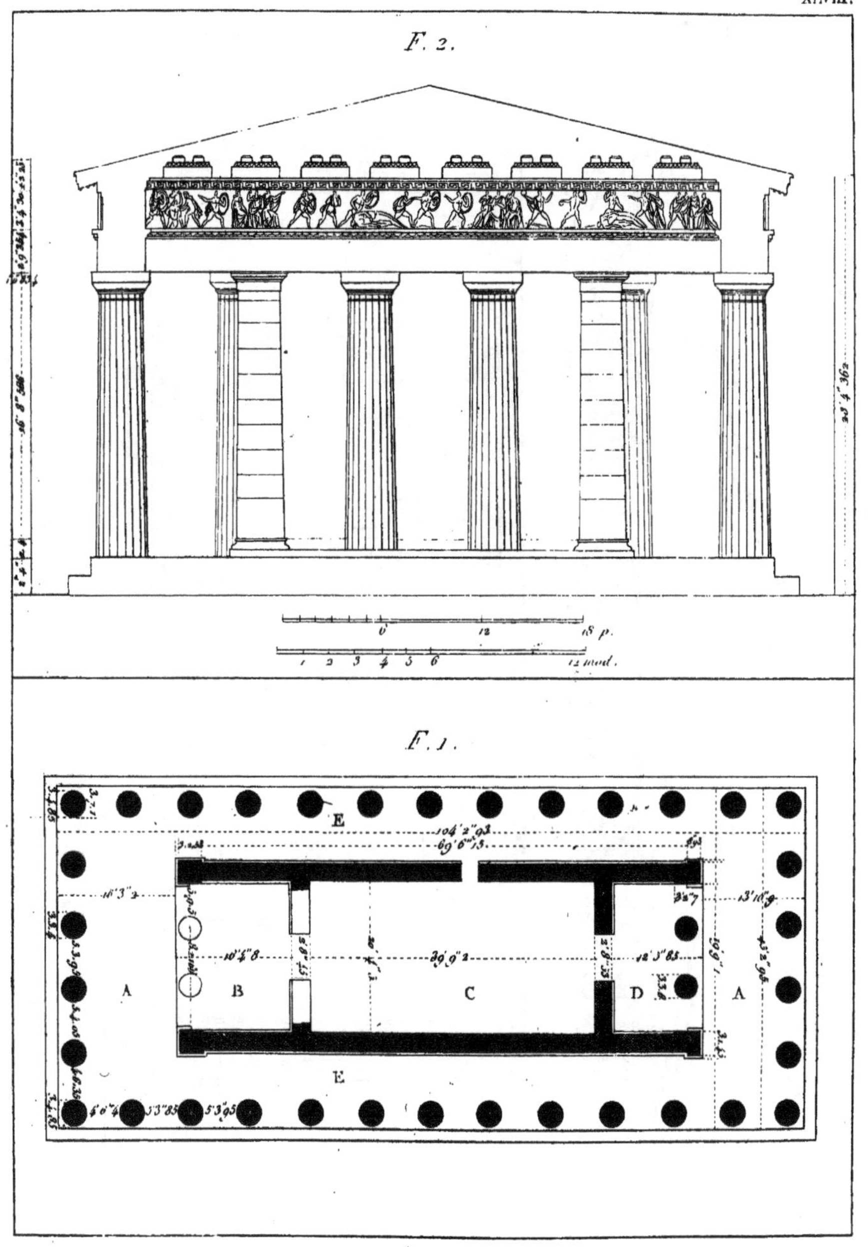
F. 2.
6
12
18 p.
1
2
3
4
5
6
12 mod.
F. 1.
E
A
B
C
D
A
E

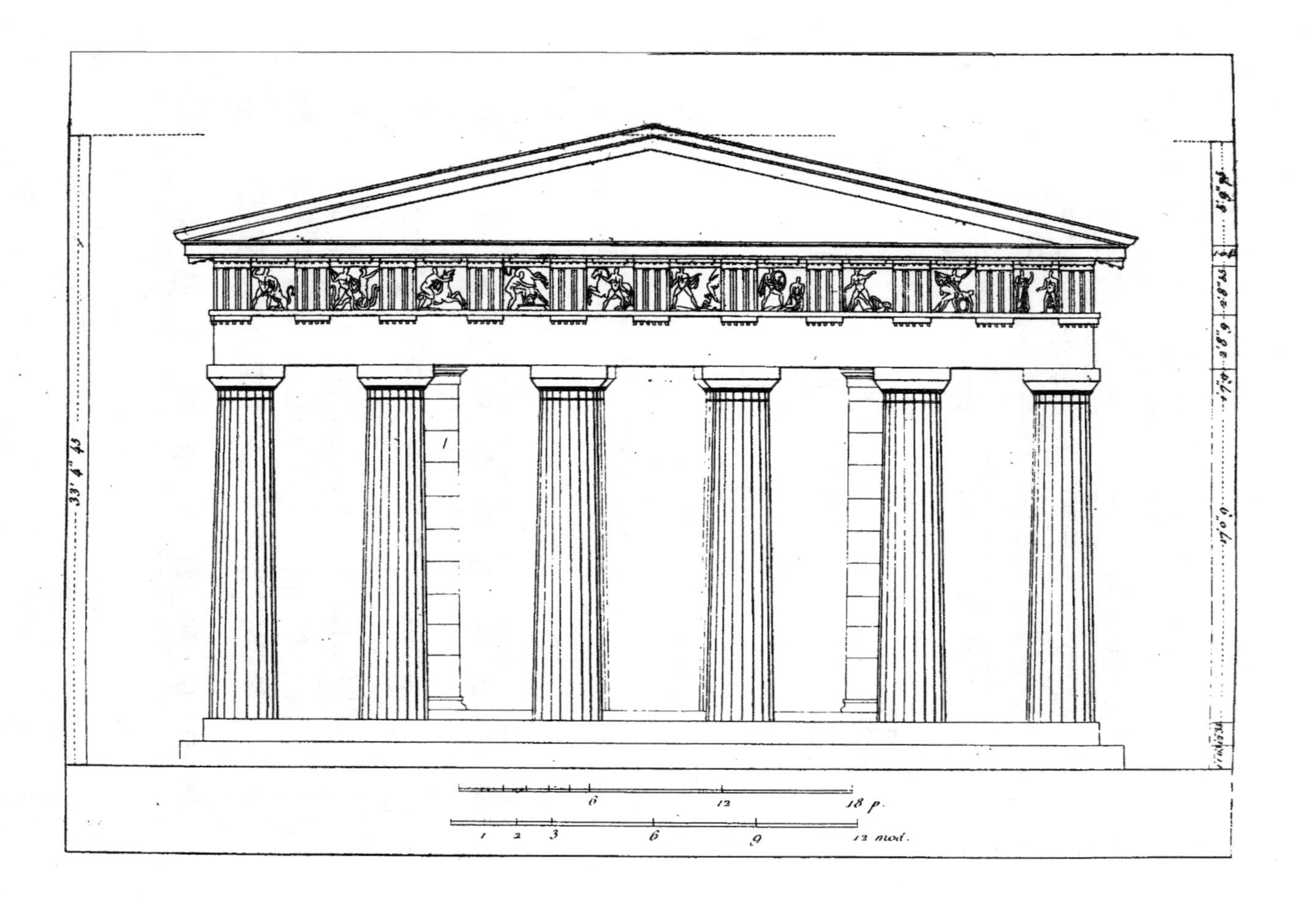
6
12
18 p.
1
2
3
6
9
12 mod.

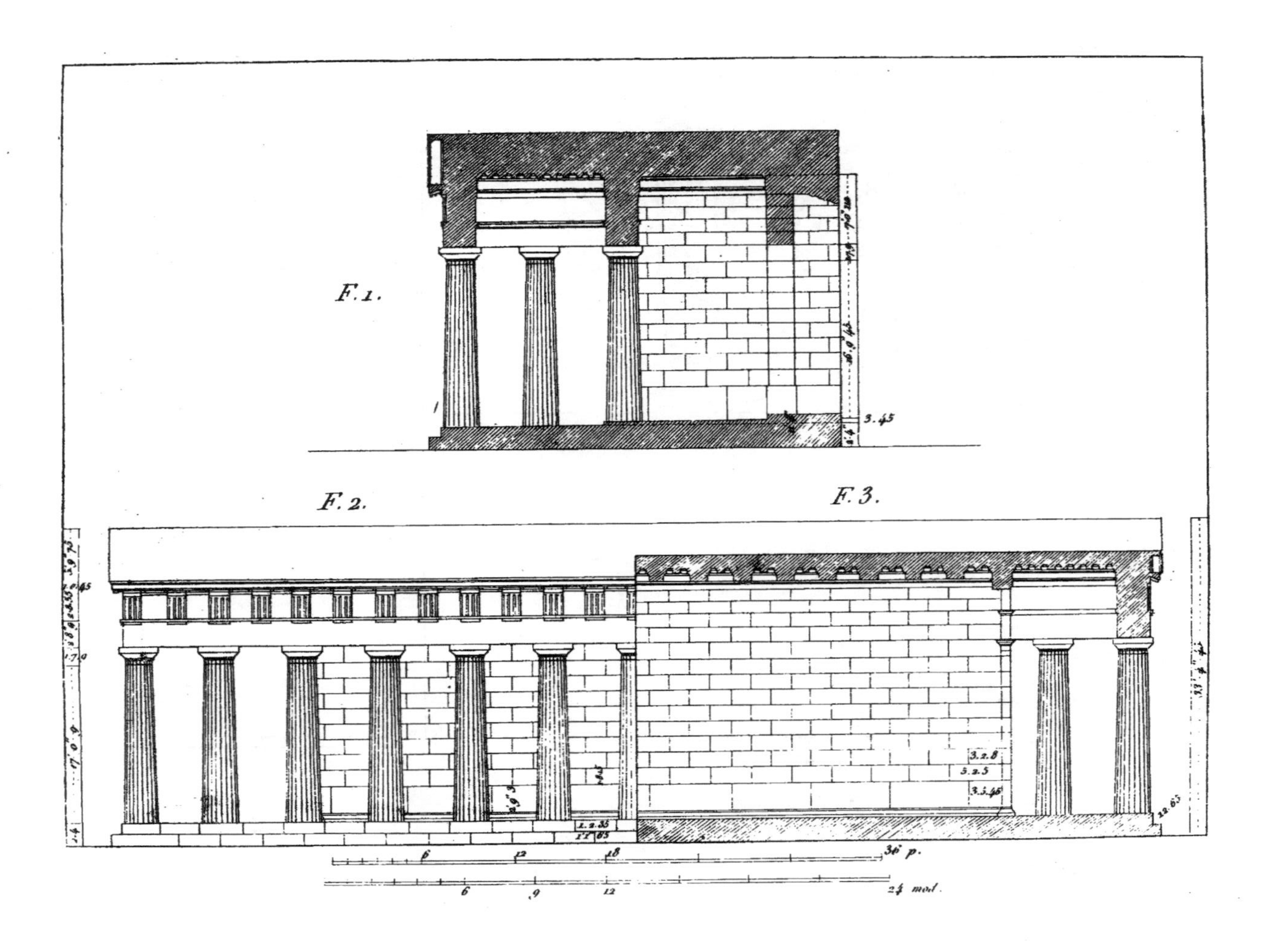
F. 1.
F. 2.
F. 3.
3.45
36 p.
24 mod.

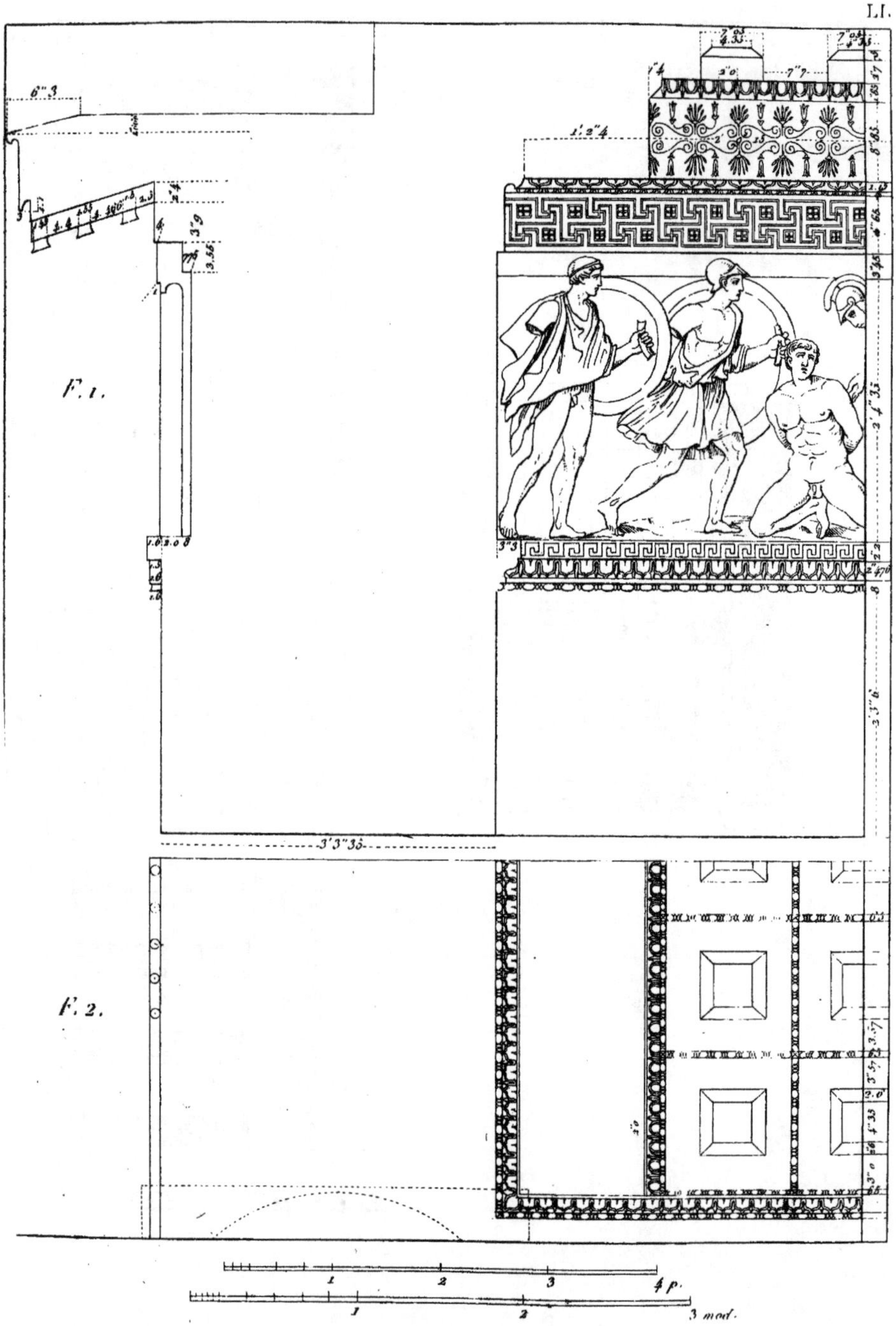

LI.

F. 1.

F. 2.

1 2 3 4 p.

1 2 3 mod.

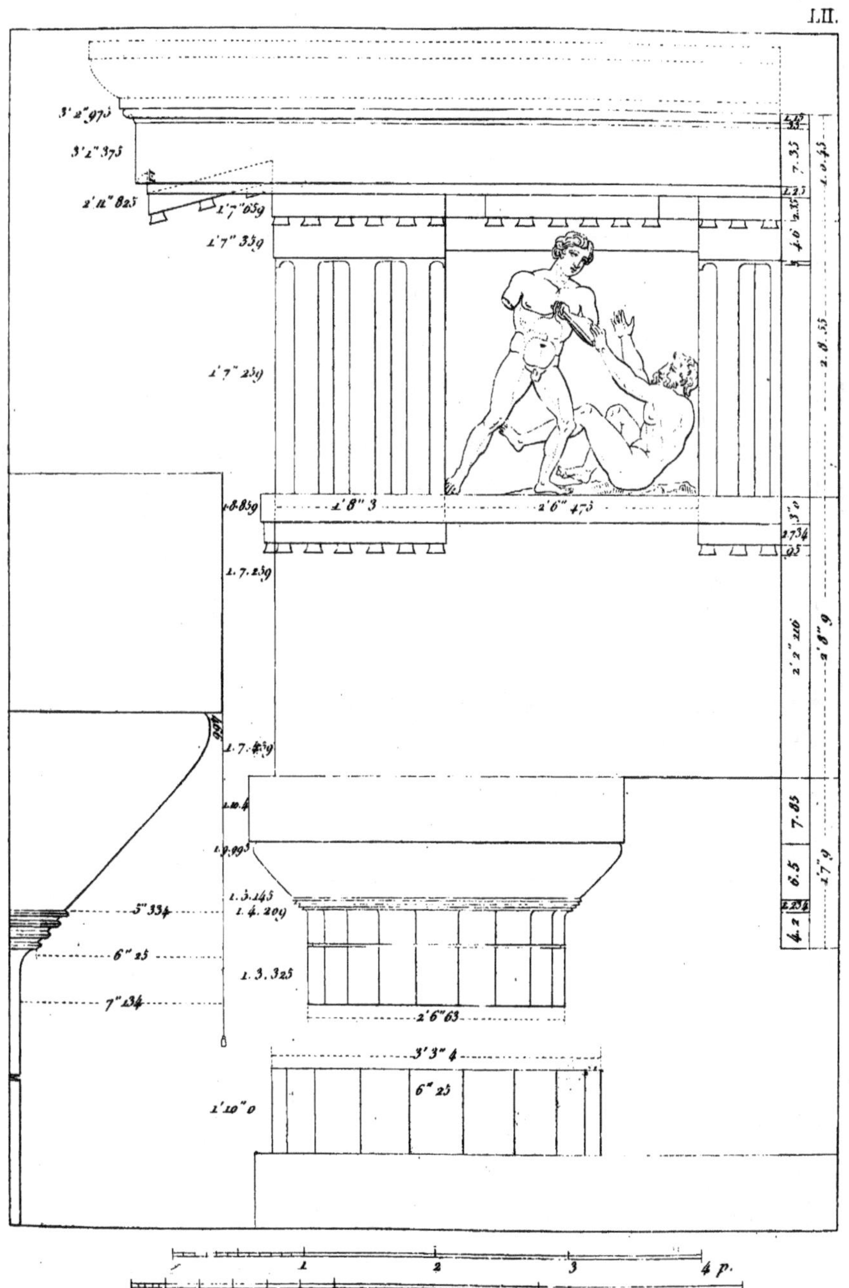

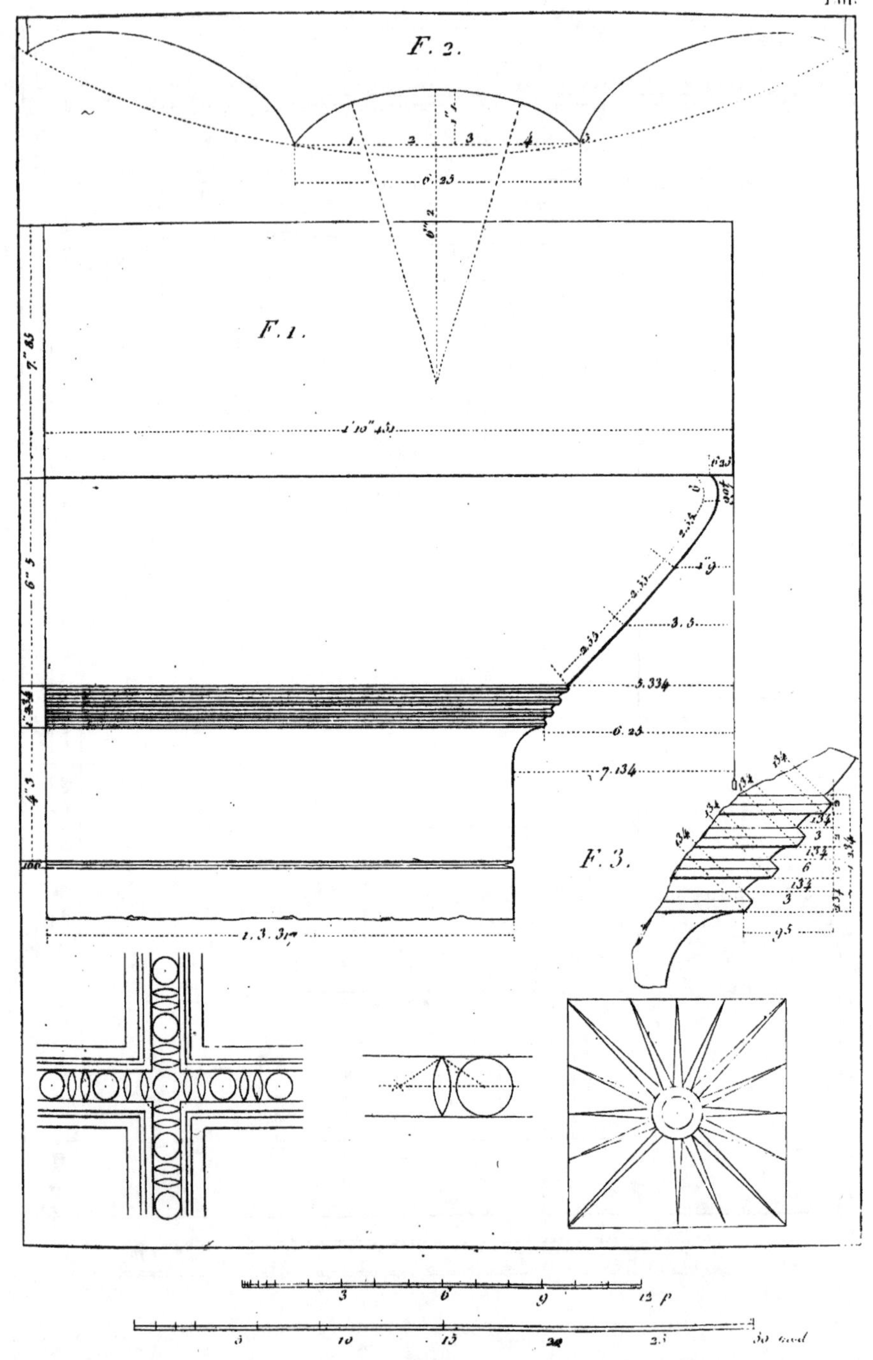
F. 2.
6. 23
F. 1.
F. 3.
3 6 9 12 p
5 10 15 20 25 30

LIV.

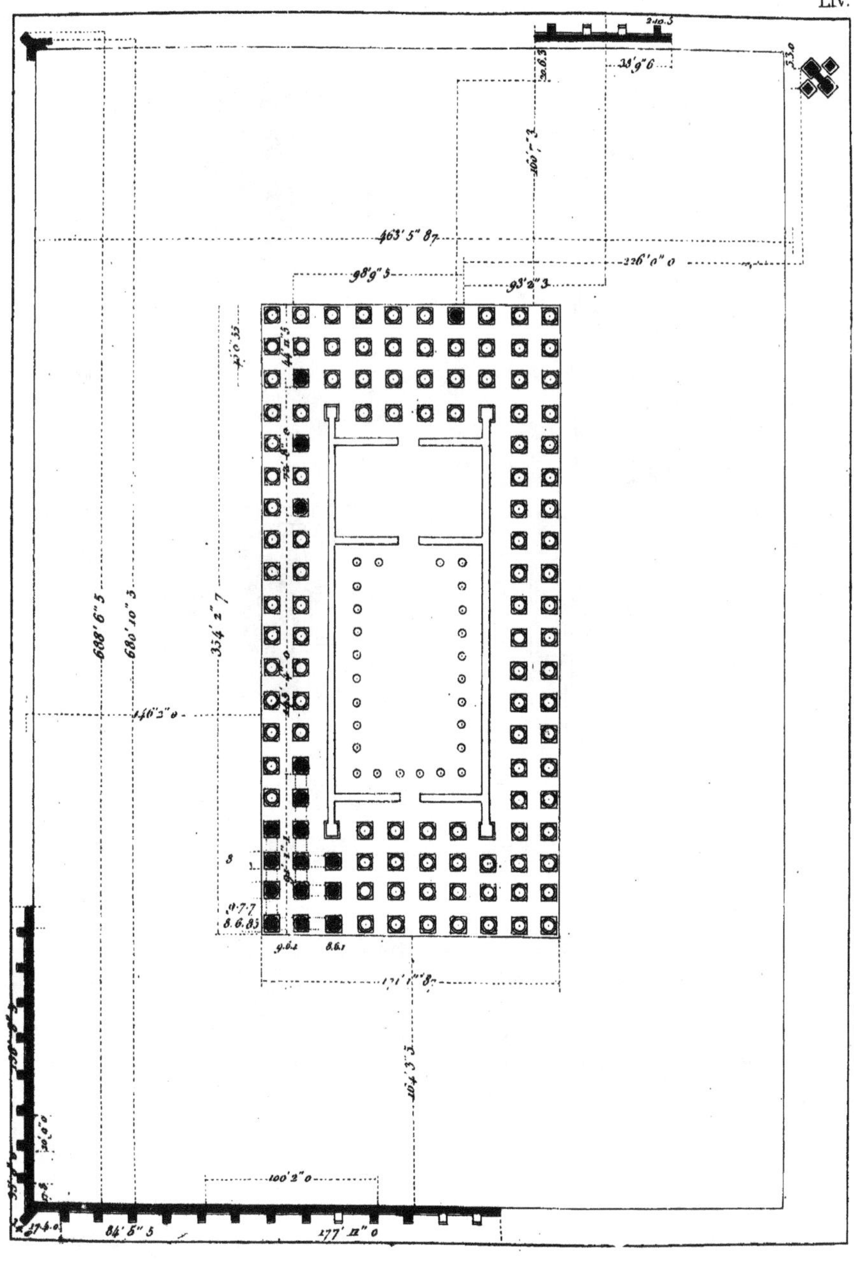
463' 5" 87
98'9" 5
93'2" 3
226'0" 0
688'6" 5
680'10" 3
334'2" 7
146'2" 0
104'3" 5
100'2" 0
84'5" 5
177'11" 0
30 p
60 mod.

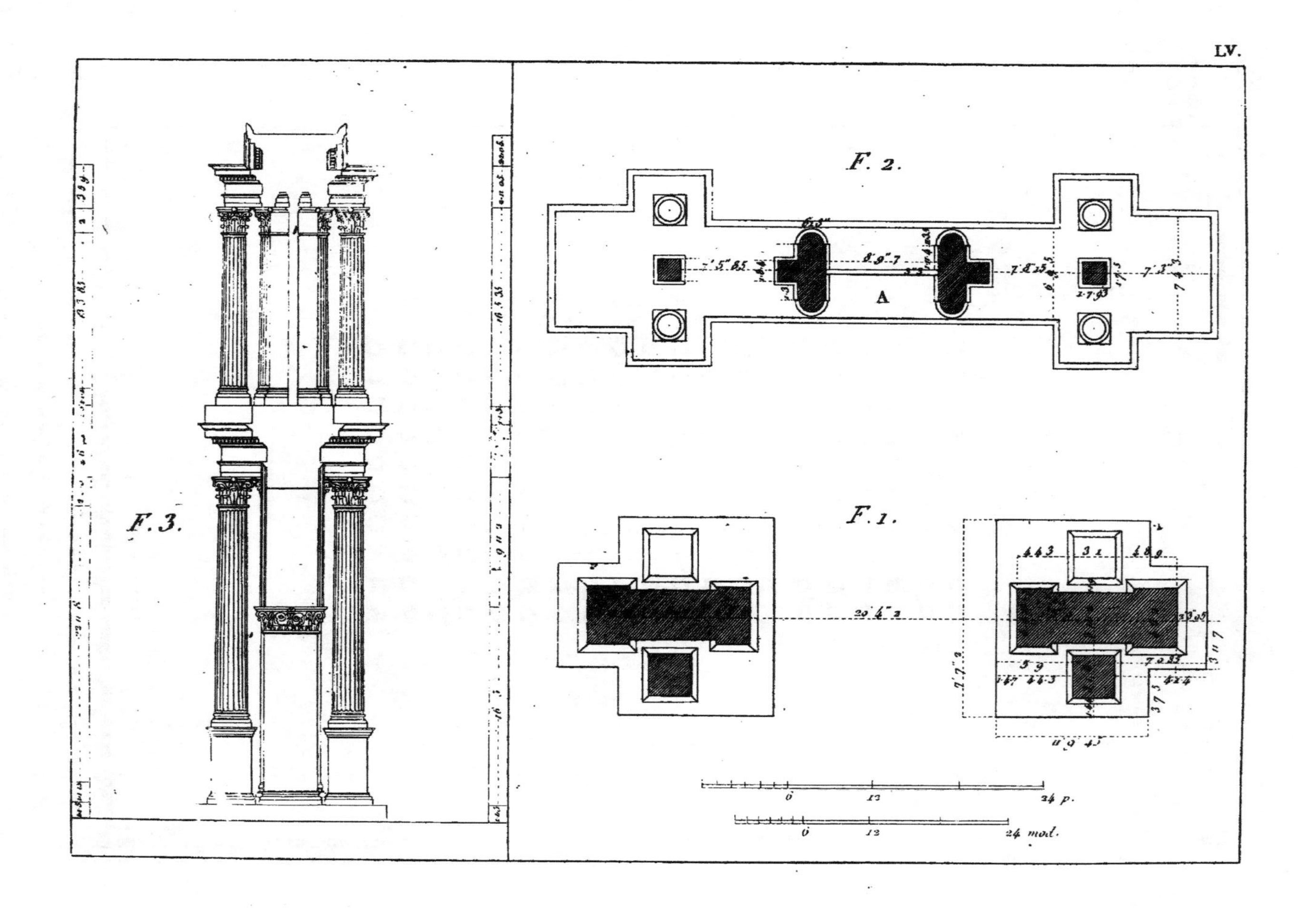
F. 2.
A
F. 1.
F. 3.
24 p.
24 mod.

6 12 18 36 p.
6 12 18 36 mod.

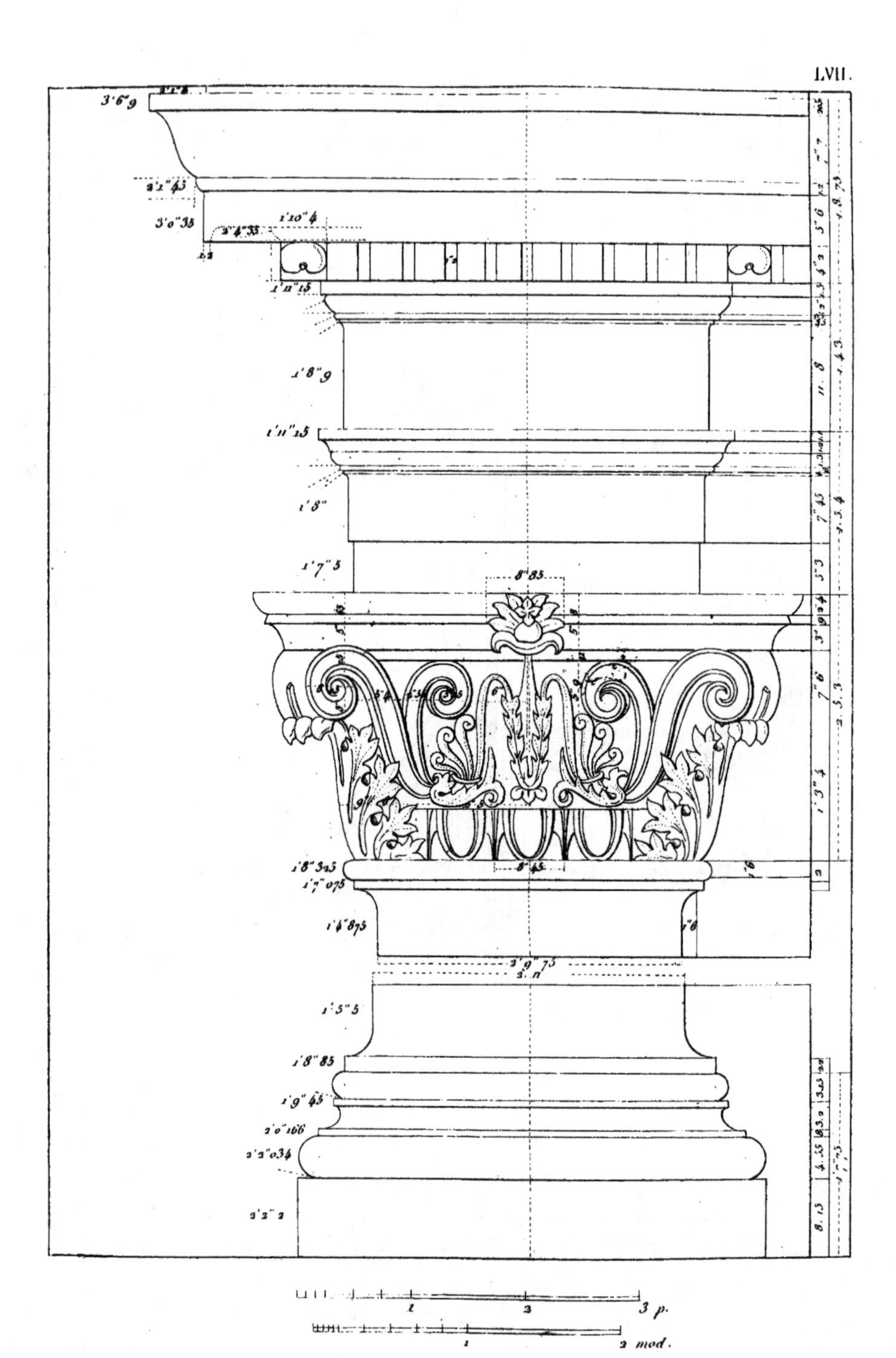

1 2 3 p.

1 2 mod.

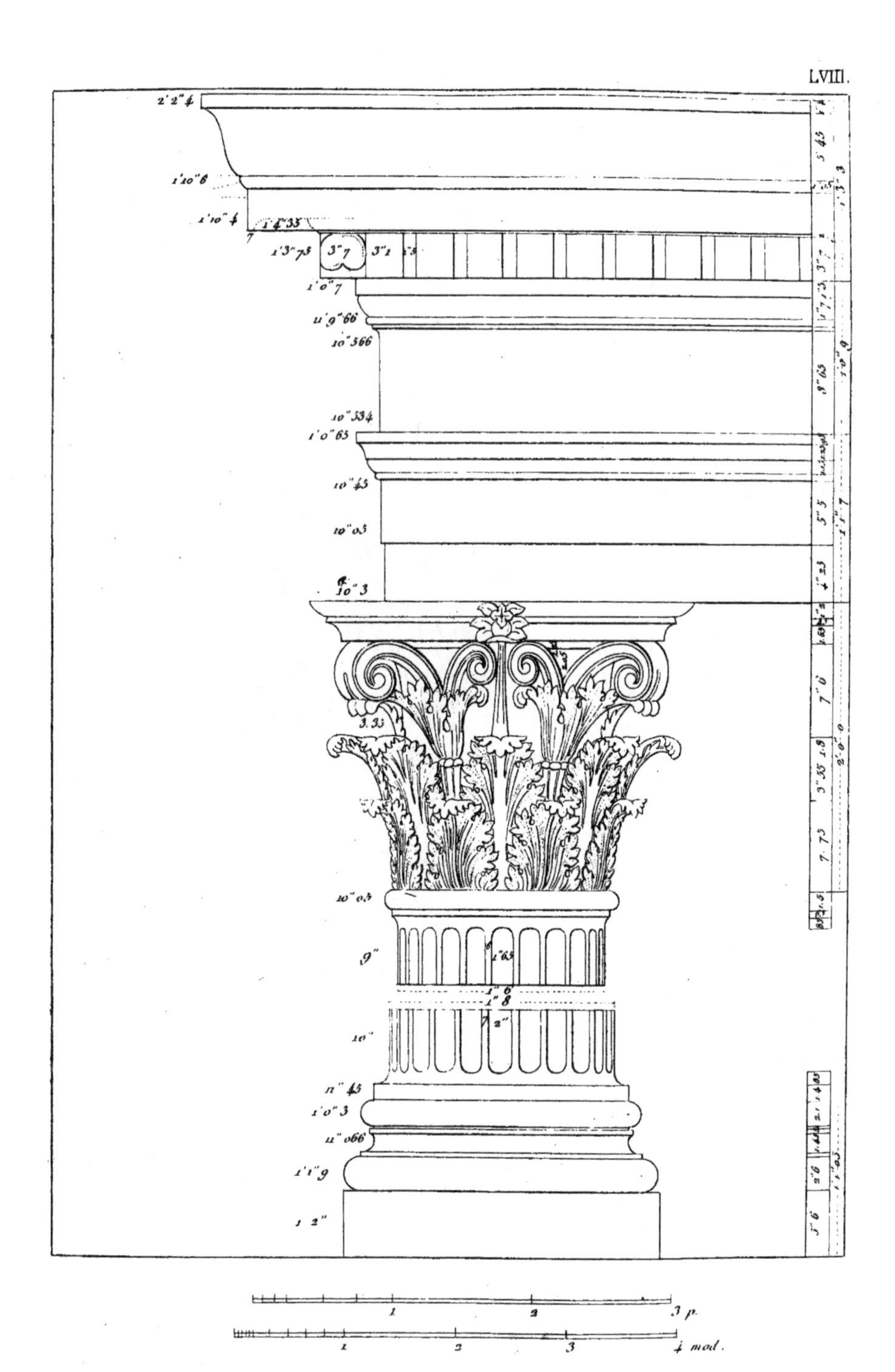

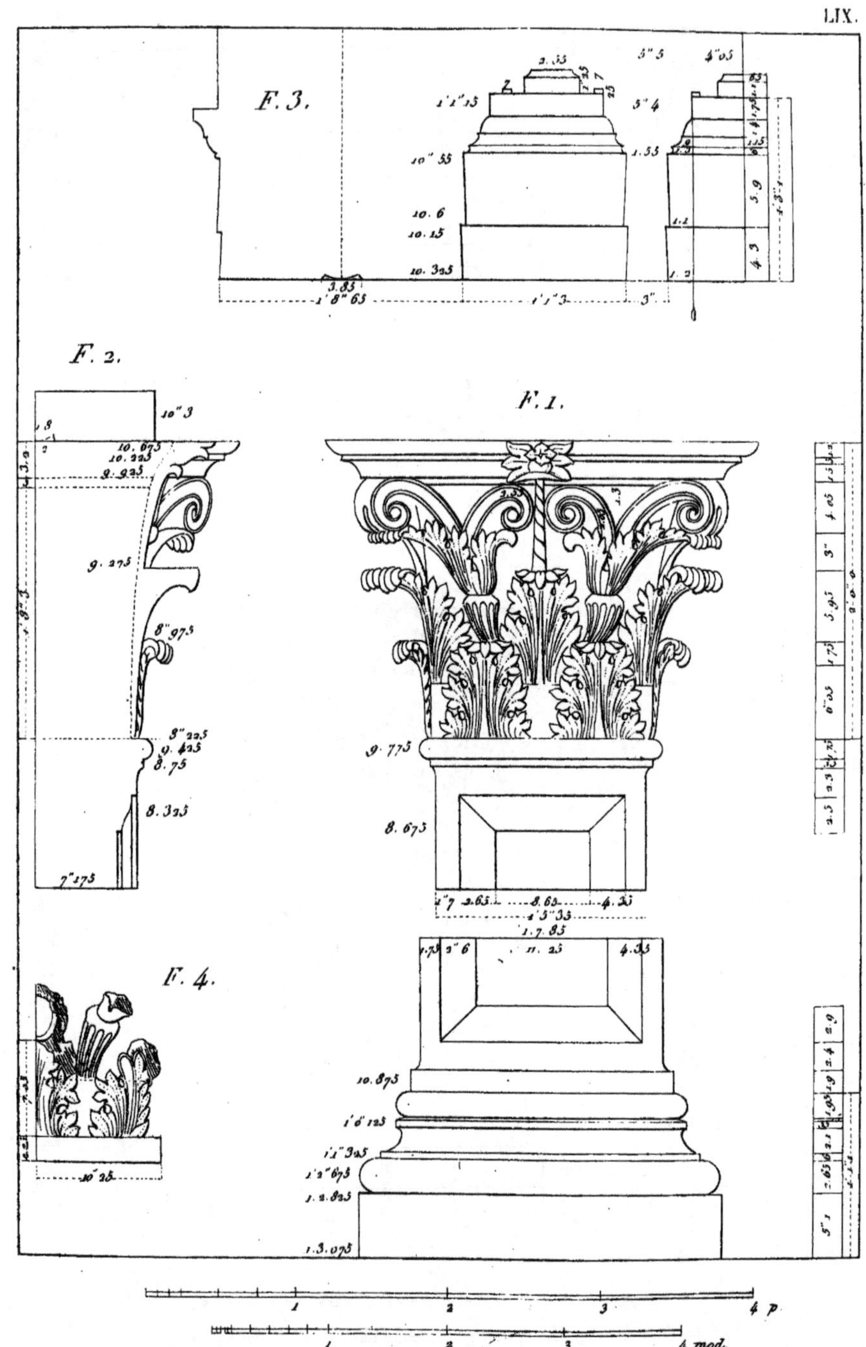
F. 3.
F. 2.
F. 1.
F. 4.
1
2
3
4 p.
1
2
3
4 mod.

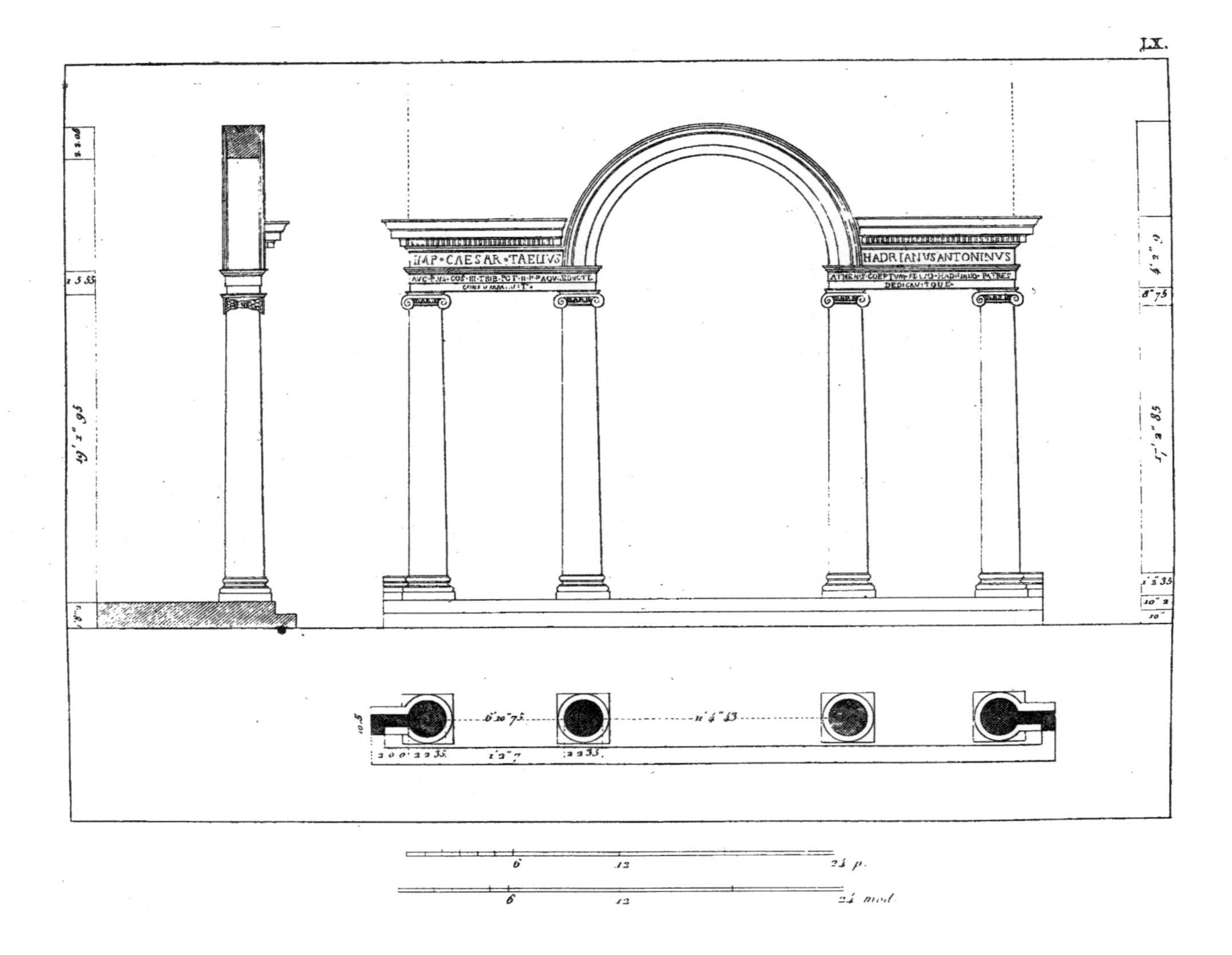
IMP·CAESAR·TAELIVS
HADRIANVS ANTONINVS
ATHENIS COEPTVM DIVO HADRIANO PATRES
DEDICAVITQVE
24 p.
24 mod.

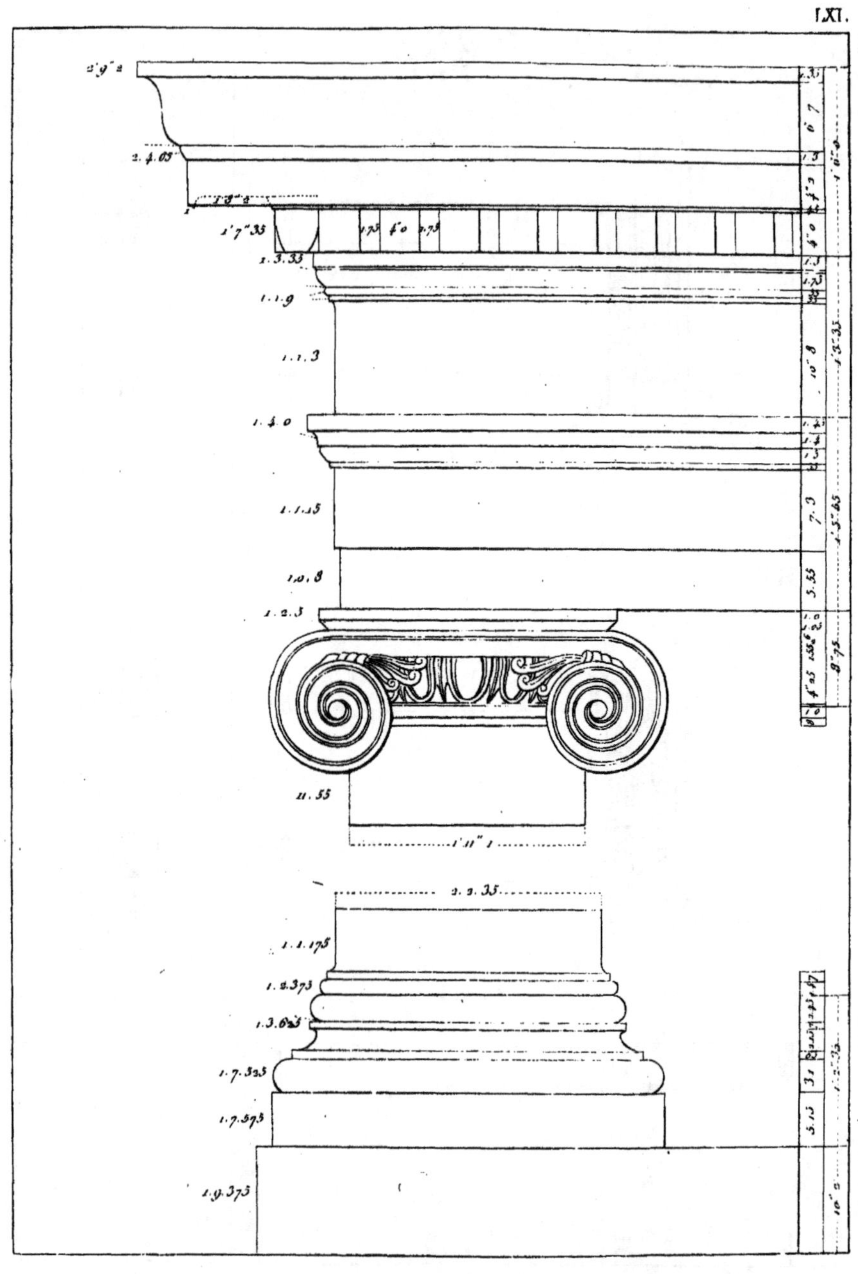
2.4.05
1'7"33
1.3.33
1.1.9
1.1.3
1.4.0
1.1.15
1.0.8
1.2.5
11.55
1'11"1
2.2.35
1.1.175
1.2.373
1.3.625
1.7.325
1.7.575
1.9.373

1 2 3 p.

1 2 3 mod.

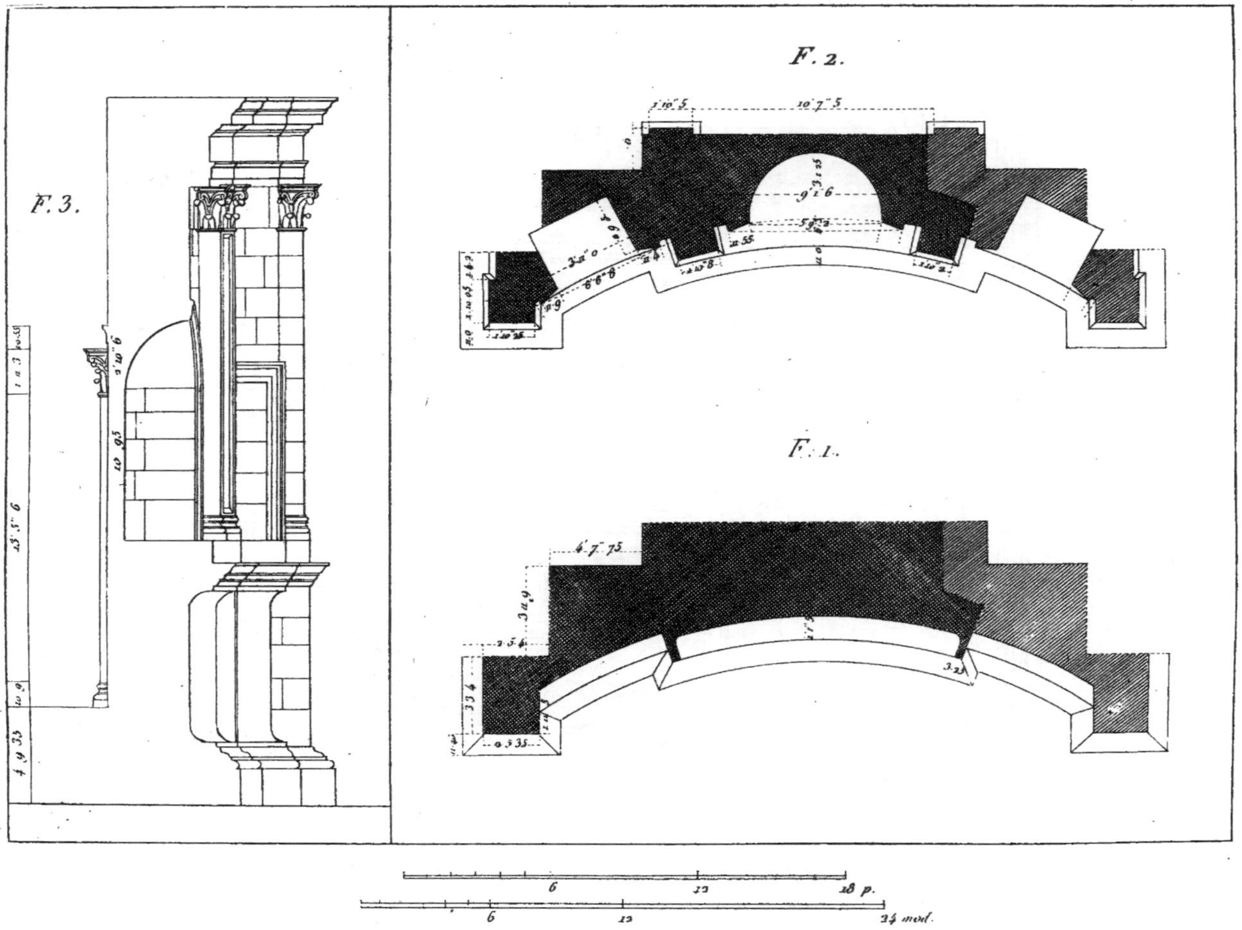

F. 3.

F. 2.

F. 1.

0 12 18 p.
6 12 18 mod.

1 2 3 p.

1 2 3 mod.

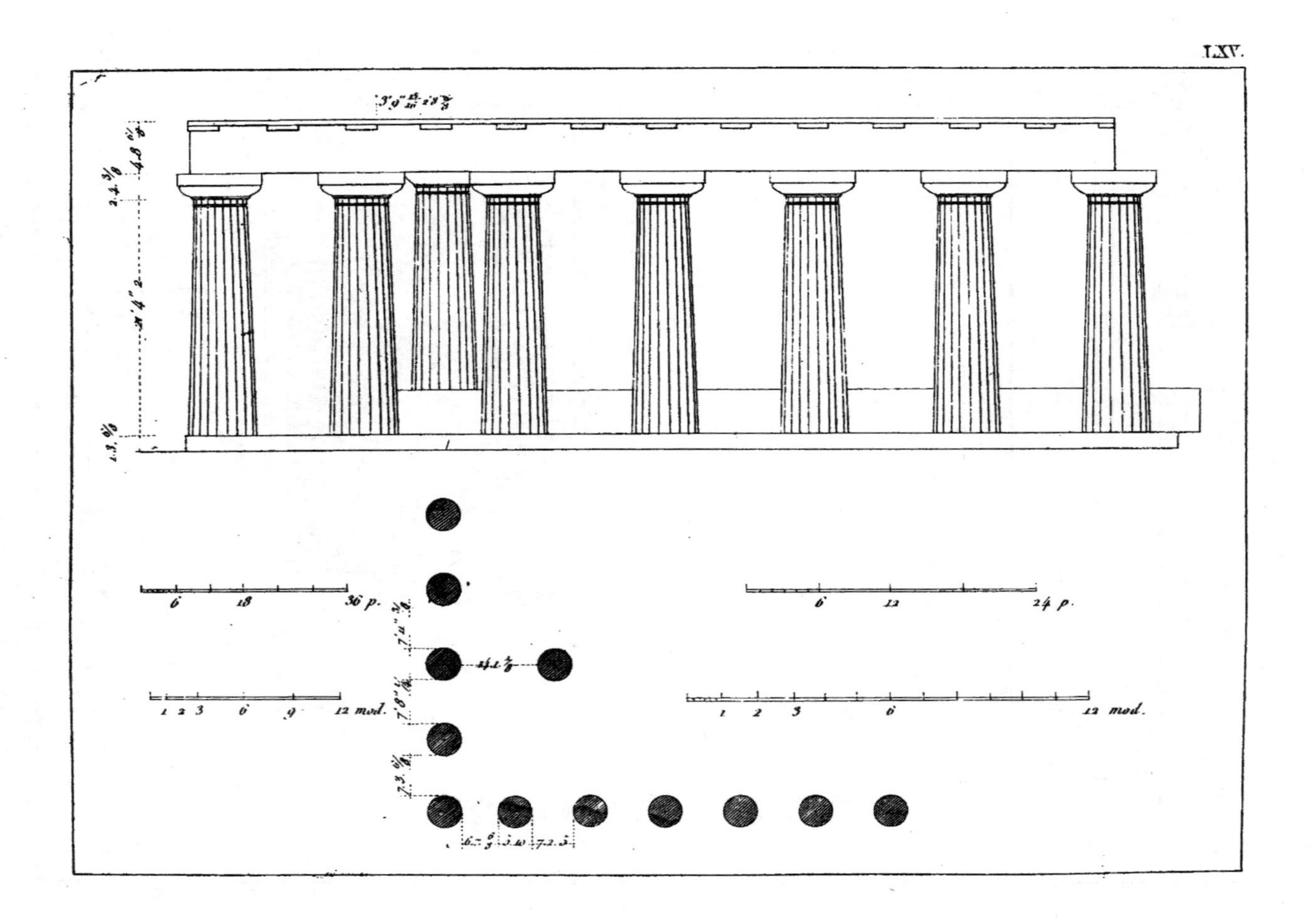
6 18 36 p.
1 2 3 6 9 12 mod.
6 12 24 p.
1 2 3 6 12 mod.

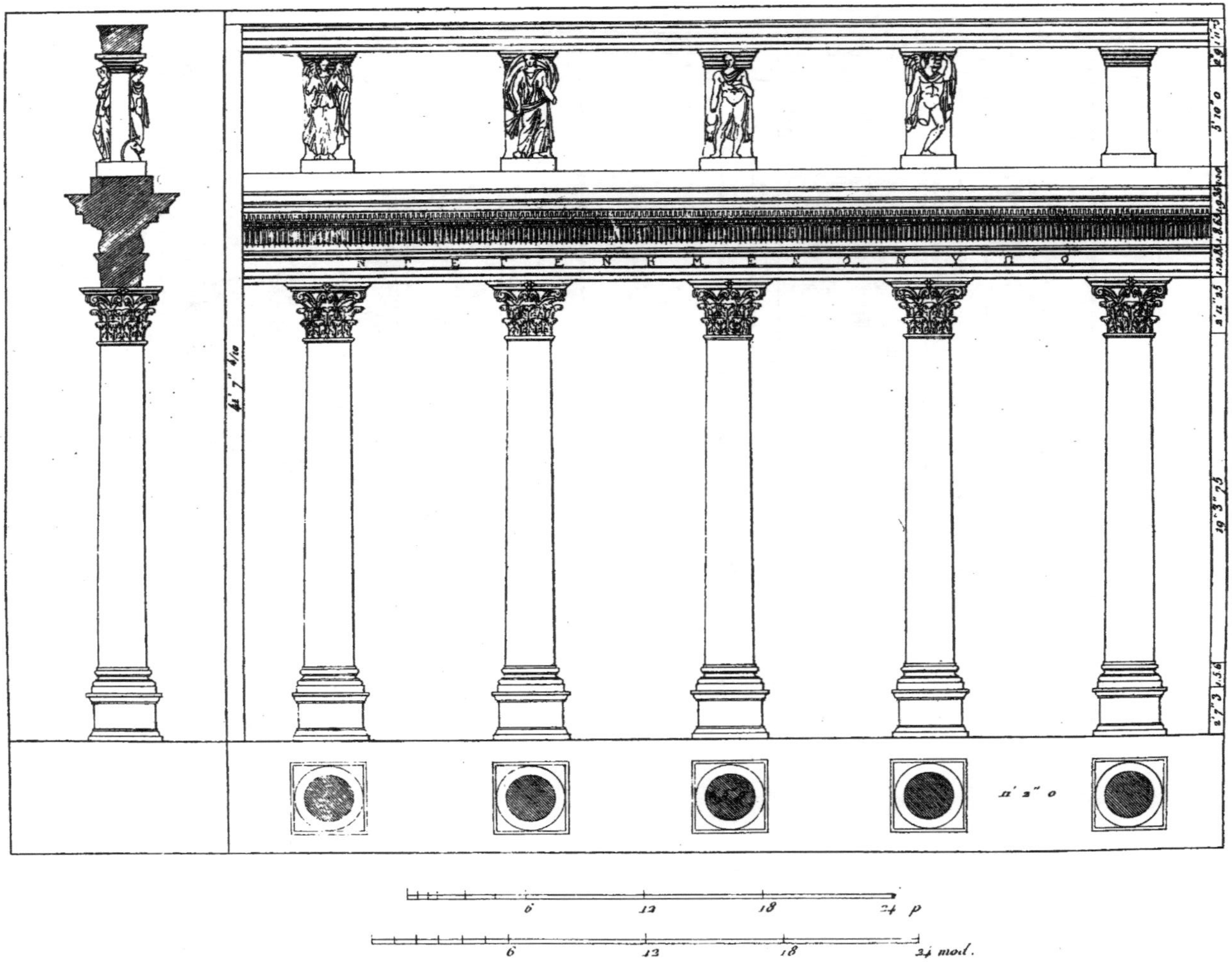

41' 7" 4/10

11' 2" 0

6 12 18 24 p

6 12 18 24 mod.

LXVII.
1
2
3
4 p.
1
2
3
4 mod.

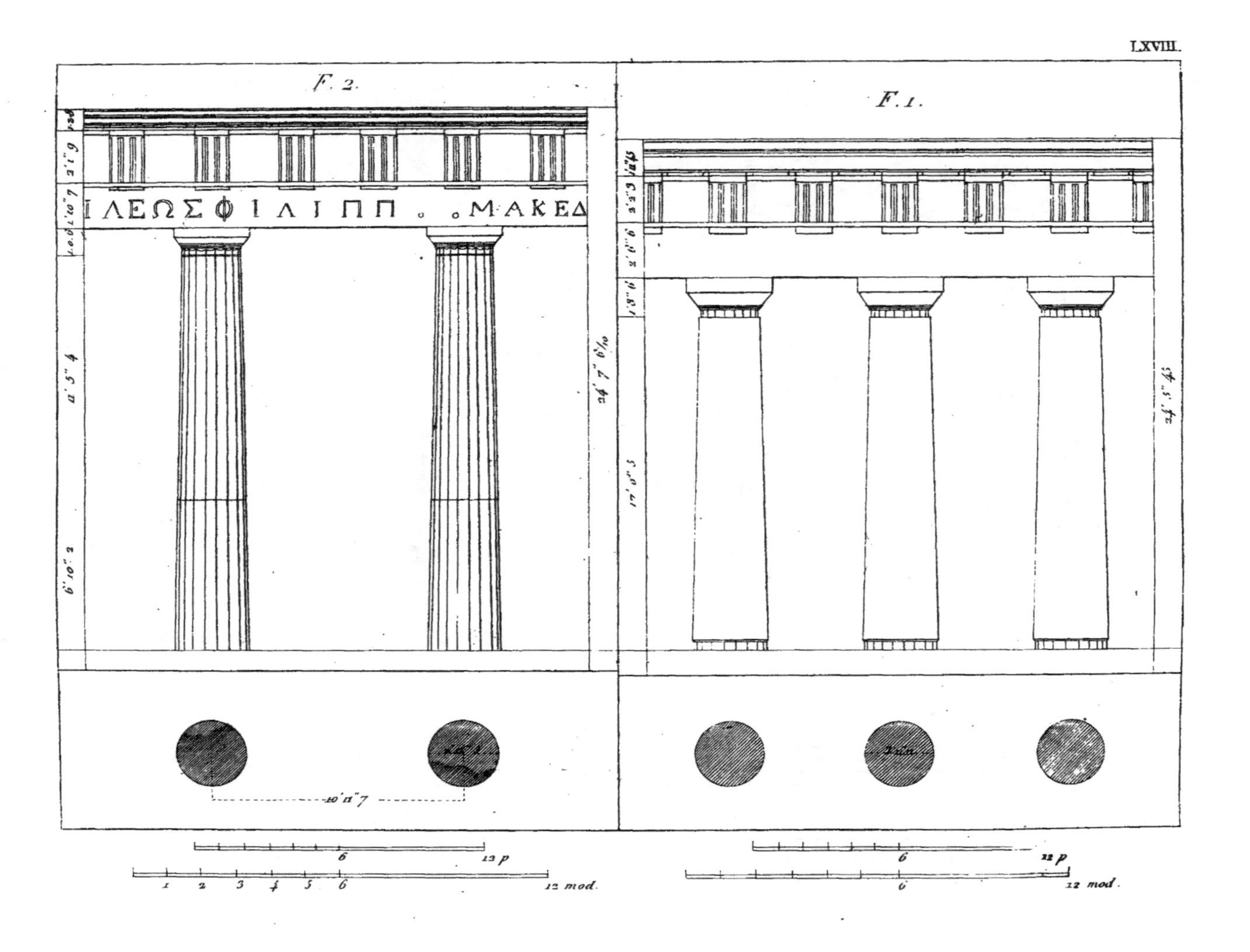
F. 2.
F. 1.
Ι ΛΕΩΣ ΦΙΛΙΠΠ . .Μ·ΑΚΕΔ
24' 7" 6/10
24' 5" 4½
11' 5" 4
6' 10" 2
17' 0" 5
10' 11" 7
12 p
12 mod.
1 2 3 4 5 6

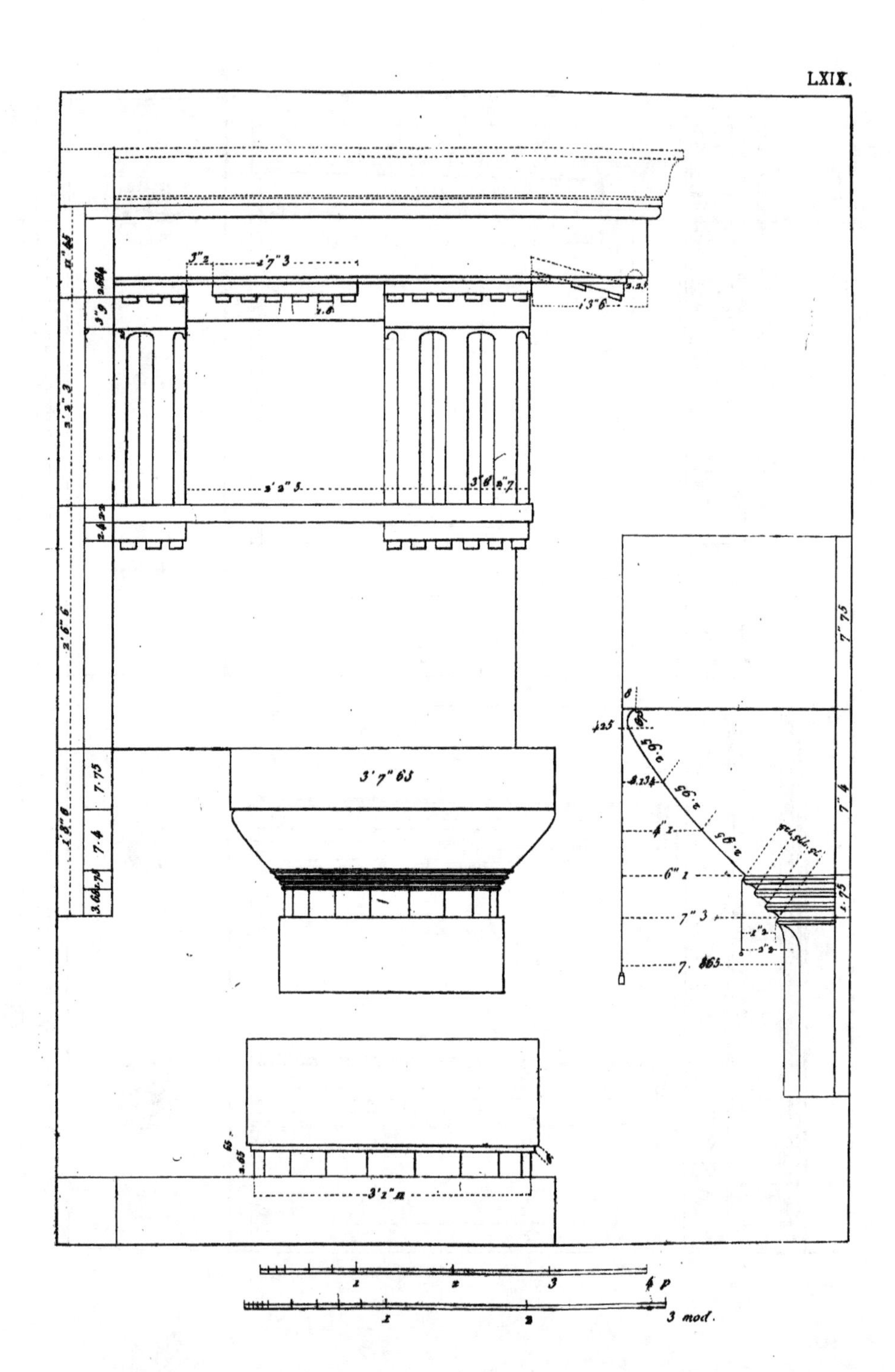
3' 7" 63
3'2" 3
1'3" 6
7.75
7.4
6" 1
7" 3
1.75
1
2
3
4 p
3 mod.

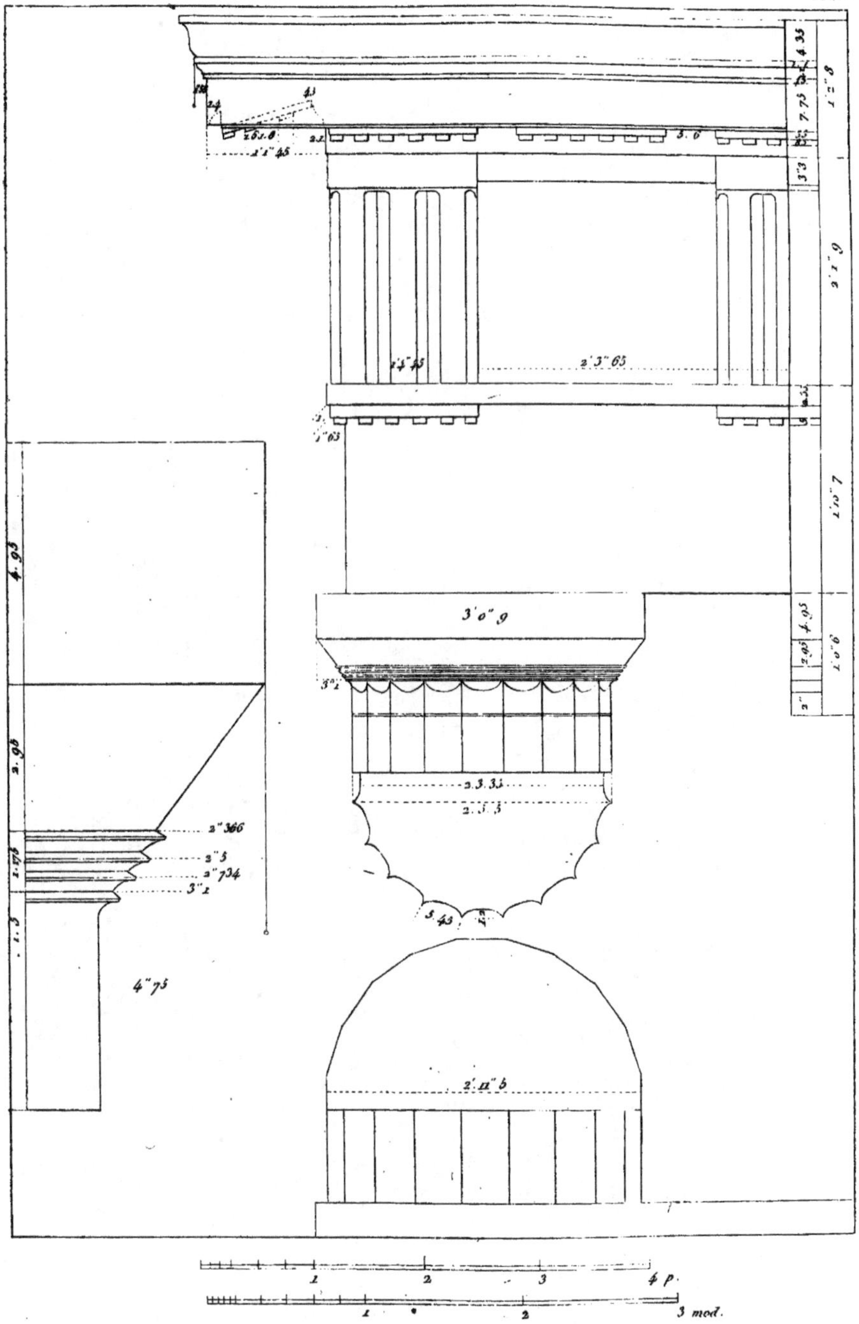

1 2 3 4 p.

1 2 3 mod.

R. Dalton delin.

A View of Mount Ætna in Sicily (now called Mon Gibello)

From lo Strozzo; with the Range of huge black Rocks which is continued along the Sea Coast for sixteen Miles. Some of these Rocks Eruptions thrown with great Violence from the Cavity of the Mountain; others have been formed from the Liquid Fiery Matter, called by ra, that issued from the Mountain in Torrents and cooling grew hard.

To John Frederick Esqr eldest Son of Sr John Frederick Bar.^t in whose presence

rliament April 12th 1751.

Vüe du Mont Ætna en Sicile (a present nommè le Mont Gibel)

Chatelain & Basire jun. sculp

Pris de l'Endroit nommè lo Strozzo, et des Amas de grands Rochers de couleur Noiratre, qui S'étendent l'Espace de Seize Miles le long de la Coste de la Mer. Partie de ces Rochers ont etè jettès dans differentes Eruptions, avec grande Violence de l'interieur de la Montagne. D'autres ont ètè formès de la Matiere Ardante & liquide (appellée par les Siciliens la Sciarra) qui etant sortie du Goufre comme un Torrent, S'est durcie et Petrifièe a mesure que la Matiere S'est rafroidie.

made; this Plate is humbly inscribed by his most obliged & obedient Servant Richard Dalton.

The Temple of Theseus at Athens, with a View of Mount Anchesmos and the adjacent

This Building is thought to be of greater Antiquity than any other at Athens being built (says Pausanias) soon after ye famous Battle of Marath

S^r. Wheler speaks of the Basso Relievos of this Temple but they are so defaced, that all one can make out from their remains is, that the Athenian Sculpt

To the R^t. Honble. the Lord Vis^t. Charlemont; This Work intended to illustrate the

Chatelain & Byfiere jun. sculp.

Le Temple de Theseé a Athénes, avec la Vüe du Mont Anchesmos et du Païs d'Alentour.

On croit que ce Batiment est le plus ancien de tous ceux d'Athénes: Car (selon Pausanias) on l'a erigé peu de tems aprés la fameuse Battaille de Marathon.

...tion.

...n & finish'd under his Lordships Direction & Patronage, is most humbly inscribed by his Lordships most oblig'd & obedient Servant R. Dalton.

Publish'd according to Act of Parliament April 12th 1751.

A View of the PARTHENION *or Temple of Minerva at Athen*

E. Rooker Sculp.

Vüe du PARTHENION ou Temple de Minerve a Athênes.

R. Dalton del. — J. Basire jun. Sculp.

A View of the PARTHENION or Temple of Minerva on the North side | Vüe du PARTHENION ou Temple de Minerve du Coté du Nord.

Publish'd according to Act of Parliament April ye 12th 1751.

The Basso Relievos on the Freze of the Inner Portico of the Temple of Minerva — I — *Les Bas Reliefs de la Frise du Portique Interieure du Temple de Minerve*

R. Dalton ad viv. fecit

The Basso Relievos on the Frize of the Inner Portico of the Temple of Minerva | Les Bas Reliefs de la Frise du Portique Interieure du Temple de Minerve

R. Dalton aqua forti fecit

The Basso Relievos on the Frize of the Inner Portico of the Temple of Minerva | *Les Bas Reliefs de la Frise du Portique Interieure du Temple Minerve*

R. Dalton aqua fortis fecit

The Basso Relievos on the Frize of the Inner Portico of the Temple of Minerva | Les Bas Reliefs de la Frise du Portique Interieure du Temple de Minerve

R. Dalton aquâ forti fecit

The Basso Relievos on the Frize of the Inner Portico of the Temple of Minerva | Le Bas Reliefs de la Frise du Portique Interieure du Temple de Minerve

R. Dalton aquâ forti fecit

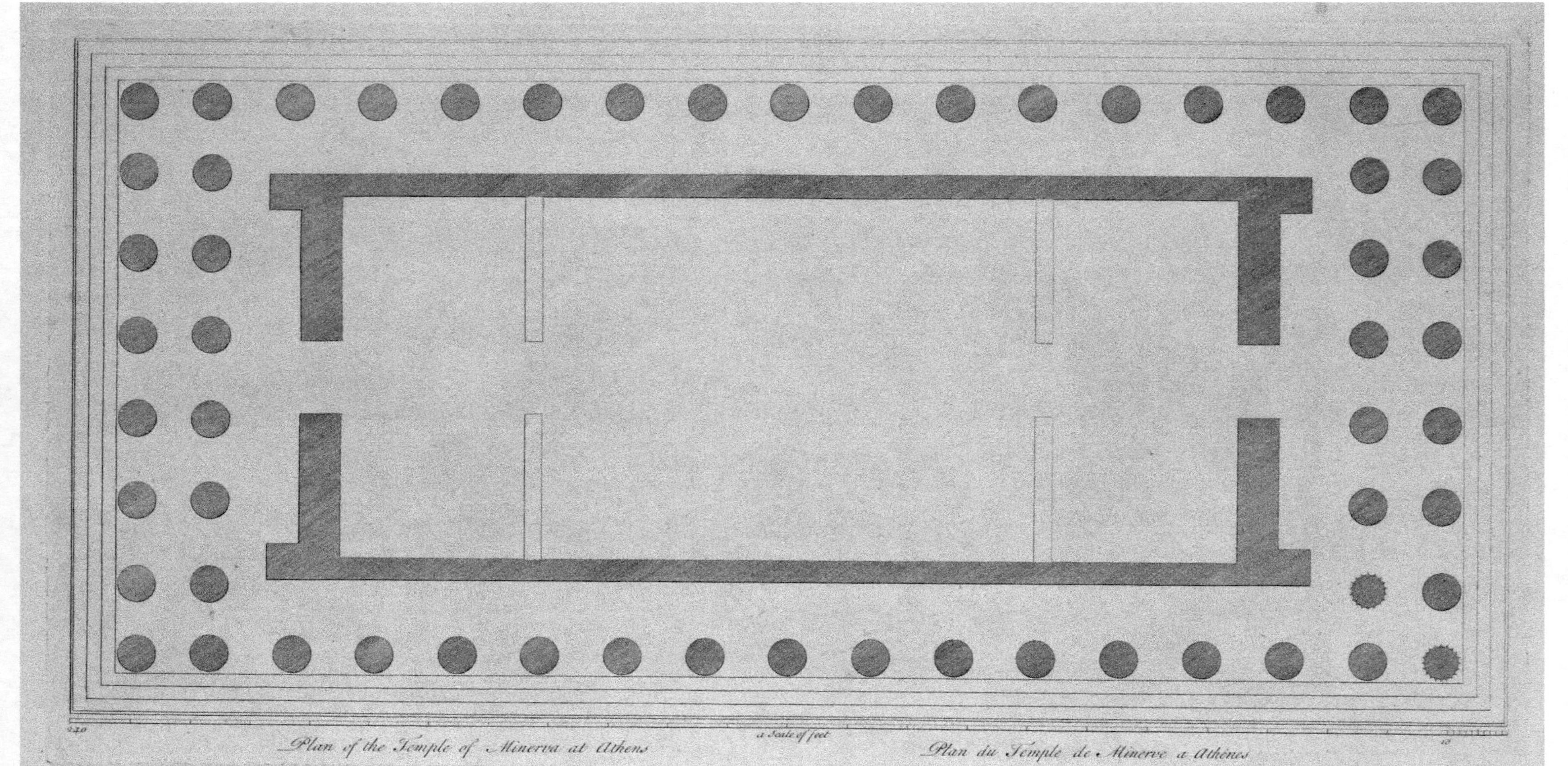

Plan of the Temple of Minerva at Athens

Plan du Temple de Minerve a Athênes

R. Dalton del.　　J. S. Müller sc.

The Temple of Erictheus at Athens.　　*Le Temple d' Erictheus a Athénes.*

Publish'd according to Act of Parliament April 1st 1751.

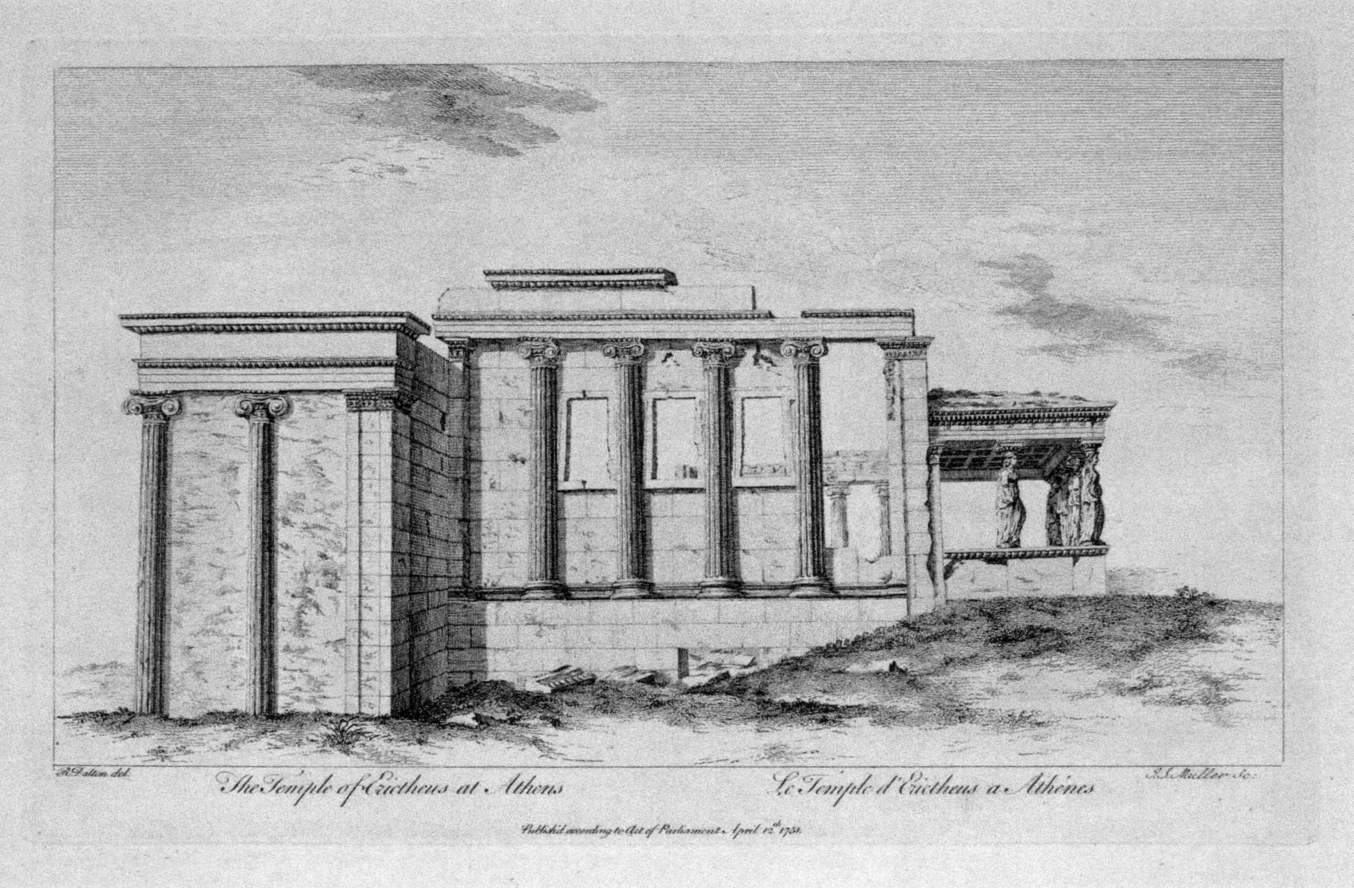

The Temple of Erictheus at Athens

Le Temple d'Erictheus a Athénes

Publish'd according to Act of Parliament April 12th 1751.

The Principal Parts of the Temple of Ericthous, in Large. — I — Les principales Parties du Temple D'Ericthée en Grand.

R. Dalton del. J.S. Müller sc.

One End of the Building contiguous to the Temple of Erictheus

Un Angle du Bâtiment contigue au Temple d'Erictheus vüe de Profil.

The Tower of Andronicus or the Temple of the Winds at Athens.

La Tour d'Andronicus autrement dite le Temple des Vents à Athénes.

Publish'd according to Act of Parliament April 12th 1751.

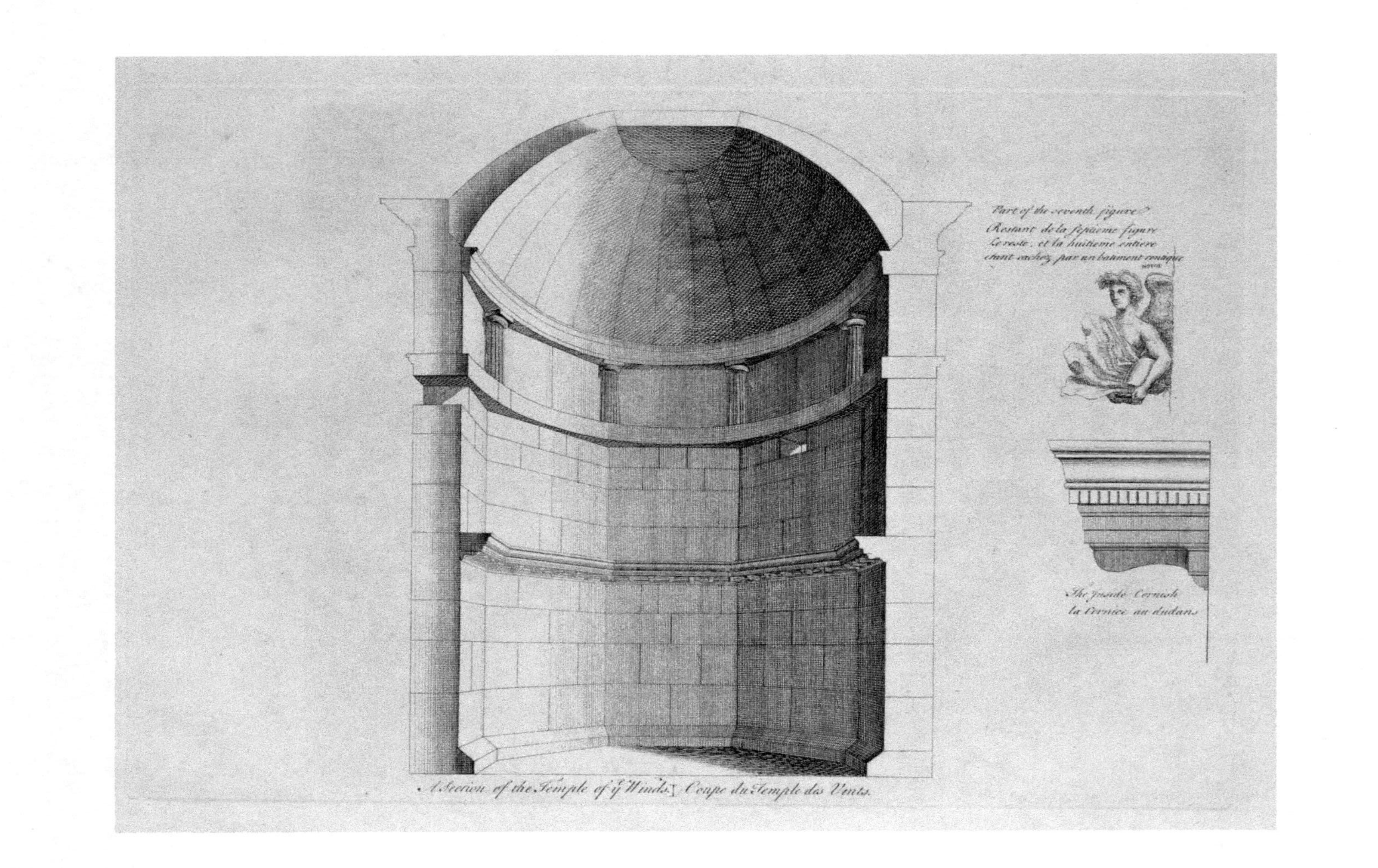

A Section of the Temple of ye Winds. Coupe du Temple des Vents.

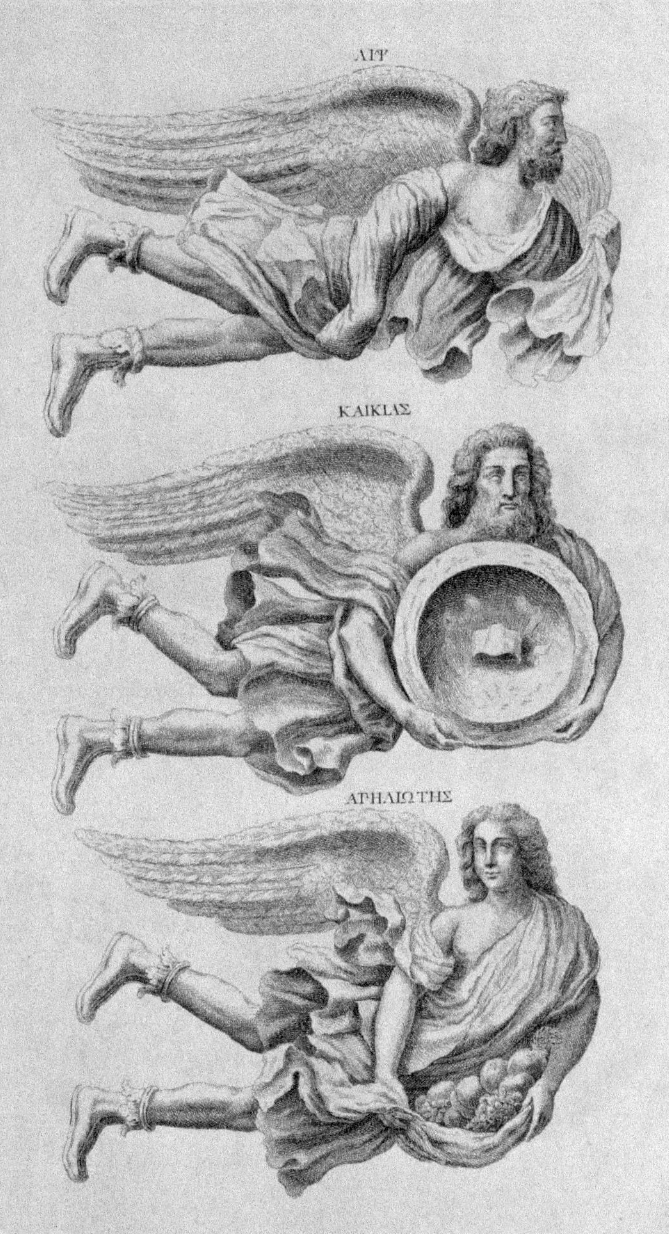

The Basso Relievos on the Freze of the Temple of the Winds. | Les Bas Reliefs de la Frise du Temple des Vents.

ΒΟΡΕΑΣ

ΣΚΙΡΟΝ

ΖΕΦΥΡΟΣ

The Basso Relievos on the Frize of the Temple of the Winds. | *Les Bas Reliefs de la Frise du Temple des Vents*

N.B. There are eight Winds represented on the Frize of this Temple, but one, and a great part of another of the Figures is intirely covered by the Wall of an adjoyning Dwelling House.

Published according to Act of Parliament Apr 1st 1751.

R. Dalton Aqua forti fecit.

R. Dalton del. A. Radigues sculp.

A Temple of Hercules commonly called the Lanthorn of Demosthenes

Temple d'Hercule dit la Lanterne de Demosthene

Publish'd according to Act of Parliament April 1st 1751

The Frize of the Lanthorn of Demosthenes.

On which are represented certain Labours of Hercules, not to be met with in any other of the Basso Relievo's now remaining.

La Frise de la Lanterne de Demosthène.

Sur la quelle on voit des Travaux d'Hercule, qu'on ne trouve pas ailleurs dans les Anciens Bas-Reliefs.

Publish'd according to Act of Parliament April ye 12th 1751.

R. Dalton aqua forti fecit

The Cupola of the Lanthorn of Demosthenes.

Le Dome de la Lanterne de Demosthène

R. Dalton delin. A. Rakgiwo sculp.

An Arch erected by the Emperor Adrian at Athens | *Un Arc erigé par l'Empereur Adrien a Athénes*

The Frize on the other side of the Arch bears this Inscription. | *La Frize de l'autre Côté de l'Arc porté cette Inscription.*

ΑΙΔΕΙΣ ΑΔΡΙΑΝΟΥ ΚΑΙΟΥΧΙ ΘΗΣΕΩΣ ΠΟΛΙΣ.

The remains of an antient Monument erected by the Athenians, on the Hill called Museum, in honour of Philopappus.

Les restes d'un Monument ancien construit par les Athéniens, sur la Montagne nommée Museum, a l'honneur de Philopappus.

R. Dalton del. J. Basire Jun. sculp.

R. Dalton del. | Mason sculp

A View of Constantinople, Pera & Galata, part of the White Sea, with the Entrance of the Bosphorus Taken from the Mountain above Scutari.

Vüe de Constantinople, Pera et Galata, partie de la Mer Blanche et l'Entrée du Bosphore : Pris de la Montagne au dessus de Scutari.

A View of the Seraglio, with part of Constantinople & the opposite Shore of Asia, viz.
the Seraglio of Scutari, part of Scutari & Leanders Tower. Taken from M.r Lisles Ho

Vüe du Serrail, avec une partie de Constantinople, et la Côte de Asie qui est vis a vis comprenant la Pointe de Calcedoine, le Serrail du Scutari, partie de Scutari et de la Tour de Leandre. Prise de la Maison de M.r Lisle, au dessus de Galata.

Publish'd according to Act of Parliament February the 15. 1752.

Entrance of the Grotto at Antiparos — l'Entrée de la Grotte d'Antiparos

The Grotto of Antiparos. (vid. Tournefort thus translated.) All these formations are of White Marble, transp
like Copper. These stems of Marble must certainly vegetate; for besides that not one single Drop of Water ever fal
of the same regularity; a Drop of Water would much rather dissipate in the fall: it is certain that none distills th
let fall a pearly Drop of Water very clear and very insipid, which no doubt was formed by the Humidity of th
We descended much lower than this place & found two Springs, one of very good fresh Water, the other Salt, rising

— To F. Pierpoint Burton Esq.r in whose prese

Publish'd according to Act of Parliam.t Feb. 12 1752.

generally break a slant & in different Beds, like the Judaic stone. Most of these pieces are covered with a white Bark, & being struck on will sound
uld not be conceivable if they did, how a few Drops falling from a great height could form cylindrical pieces, terminating like round caps, & always
to, as it does into common subterranean Cavities. All that we could find here of this nature, was some few indented sheets of stone, the points of which
h a place must condense into Water, as it does in Apartments lined with Marble.) (thus far from Tournefort)
f each other. Mons.r Tournefort says this Cavern is 25 or 30 fathom high, but it is not 20 as may be seen by the figures which are drawn in true proportion to the rest.
was made: This Plate is humbly inscribed, by his most obliged Servant. Richard Dalton

R.d Dalton delin.t J. Basire Jun.r Sculp.t

A View of the Pyramids of Gize, formerly Memphis.

To the Rt. Honble. the Lord Viscr. Charlemont, This Work intended to illustrate the Antiquities of Egypt, undertaken & finishe

Vue des Pyramides du Gize anciennement Memphis.

his Direction & Patronage, is most humbly inscribed by his Lordship's most obliged and obedient Servant R. Dalton.

Publish'd according to Act of Parliament February 1752.

A View of the two great Pyramids of Gize on the North side where the Situation of the Entrance is seen of the greatest Pyramid

Veüe des Pyramides du Gize au coté du Nord, au on voit la situation de l'Entrée de la grande Pyramide

The broken part about the Entrance of the great Pyramid of Gize. *Portion brisé autour de l'Entrée de la grande Pyramide du Gize.*

R. Dalton del. I. Mason sculp.

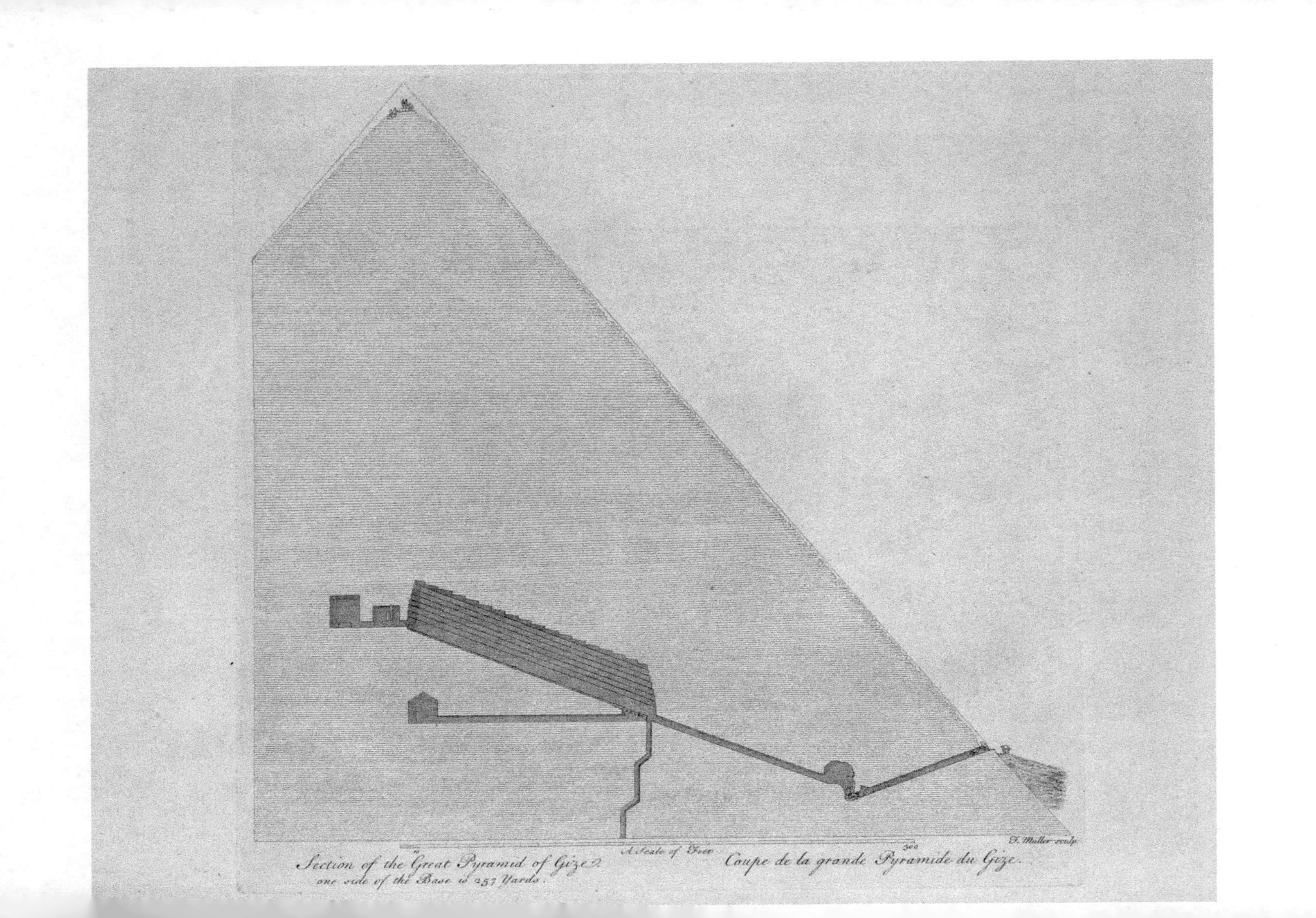

Section of the Great Pyramid of Gize.
one side of the Base is 257 Yards.

Coupe de la grande Pyramide du Gize.

A View of the Sloping Gallery
Vue de la Galerie du Pyramide du Gize

Parts whose situation are found in the Section of the whole Pyramid of Gize
Parts qui ses trouvent situé dans la coupe du Pyramide

Par les lettres F.I. s'exprime (Feet & Inches) c'est à dire, Pieds et pouces d'Angleterre

Lesser Chamber of soft free stone like the rest of the whole building, its length is 18 F. 10 1/2 I., broad 17 F. 1/2 I.

La plus petite Chambre de pierre de taille comme le reste de l'Edifice

R. Dalton del. J. Miller sculp.

The largest Chamber of the Pyramid of Gize all of Granite with the situation of the Toomb. length of the Chamber 34 feet 4 Inches & ½, breadth 17 feet 2 Inches, height 19 feet. The large Stone over the entrance is 10 feet 3 Inches long 7 feet 10 Inches broad. The Toomb is 7 feet 5 Inches & ¼ long, broad 3 feet 2 Inches & ½, high 3 feet 4 Inches, thickness of the sides 6 Inches. As the whole Room is coverd with Nine Stones, in this Chamber the great nicety of Egyptian Masonry may be observ'd.

La plus grande Chambre du Pyramide de Gize, toute de Marbre Granite, avec la situation du Tombeau comme toute la Chambre est couverte avec neuf pierres, on y peut bien observer la grande delicatesse de la Maçonnerie Egyptienne.

B. Baldan del.   J. Müller sculp.

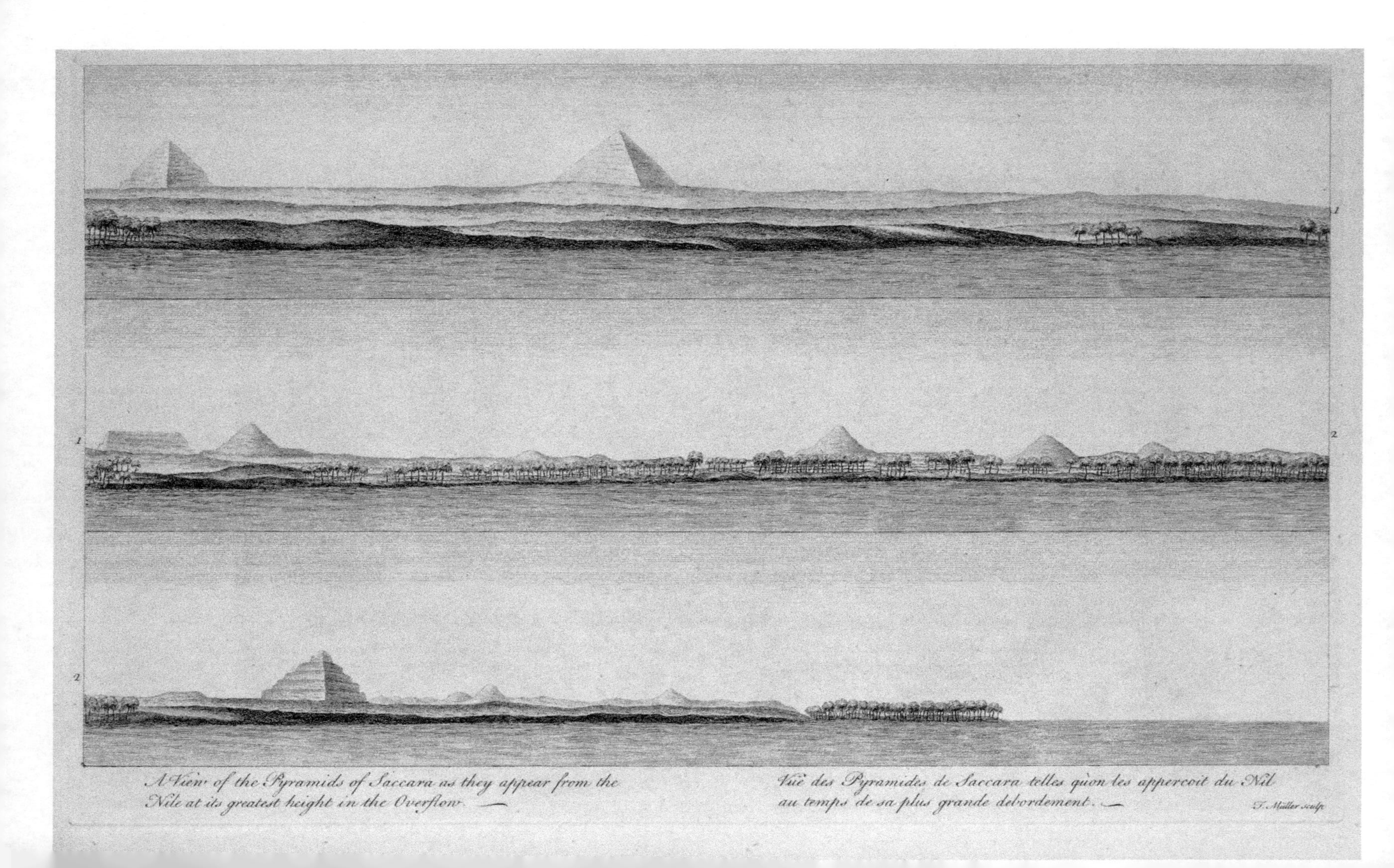

A View of the Pyramids of Saccara as they appear from the Nile at its greatest height in the Overflow.

Vüe des Pyramides de Saccara telles qu'on les appercoit du Nil au temps de sa plus grande debordement.

A View of two Pyramids at Saccara,
the largest is open.

Vue des deux Pyramides a Saccara,
dont la plus grande est ouverte.

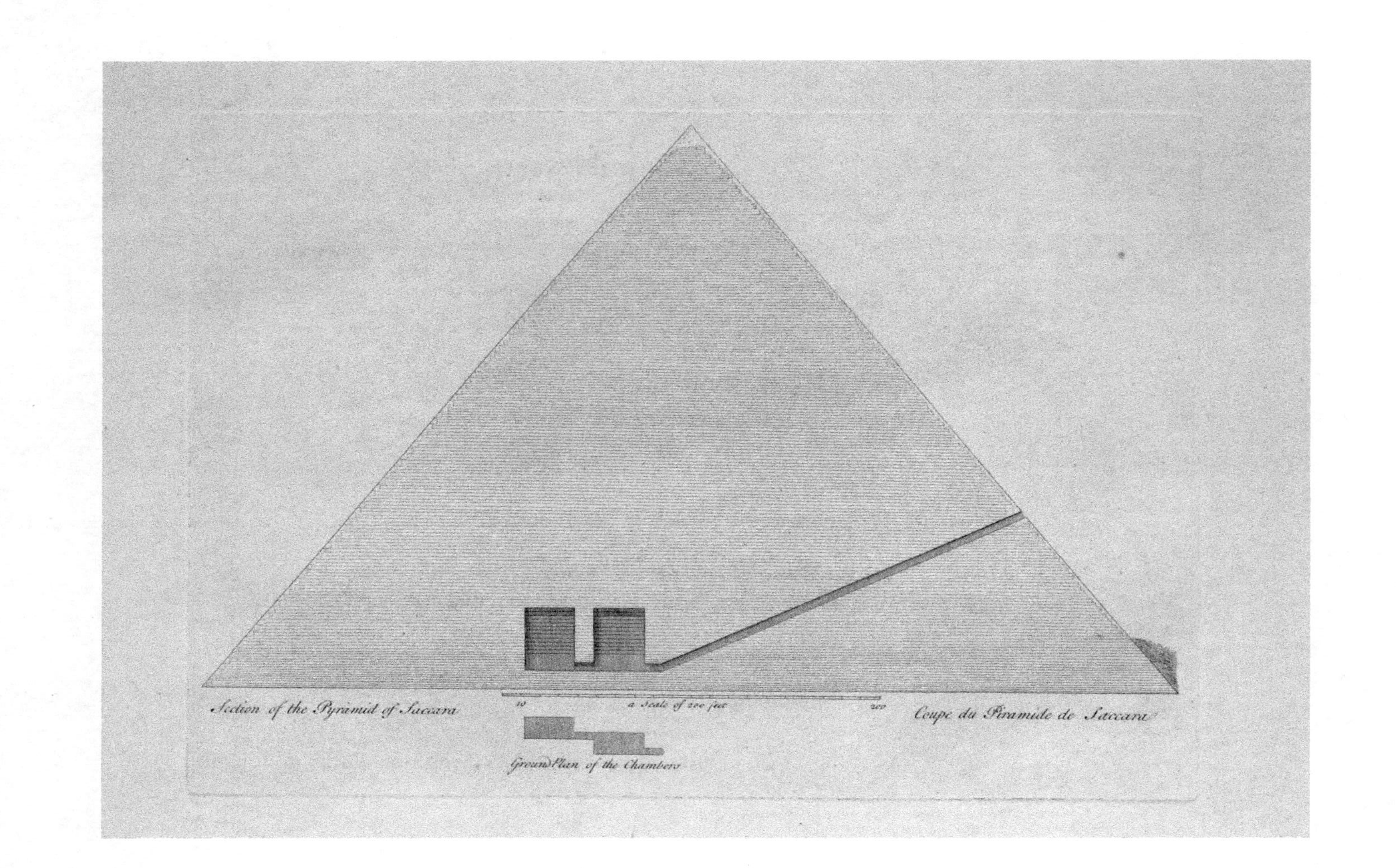

Section of the Pyramid of Saccara

Coupe du Piramide de Saccara

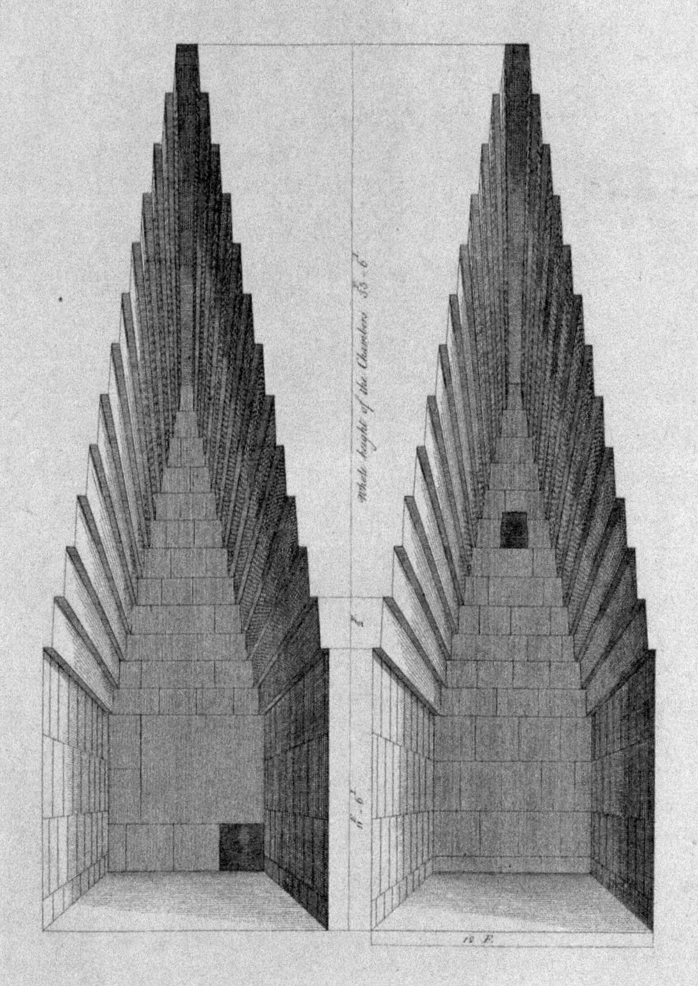

Chambers of the Pyramid at Saccara, long 18. 10 ½, broad 7. ½
large Stone over the entrance 9. 6 broad 7. ½

Chambres de la Pyramide a Saccara

J. Basire sculp.

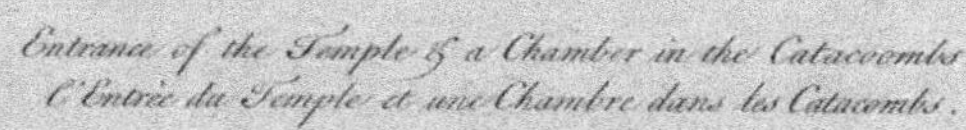

Entrance of the Temple & a Chamber in the Catacoombs.
L'Entrée du Temple et une Chambre dans les Catacombs.

H. Dalton delin. J. Basire sculpt

A Temple in the Catacoombs at Alexandria
un Temple dans les Catacombs d'Alexandrie

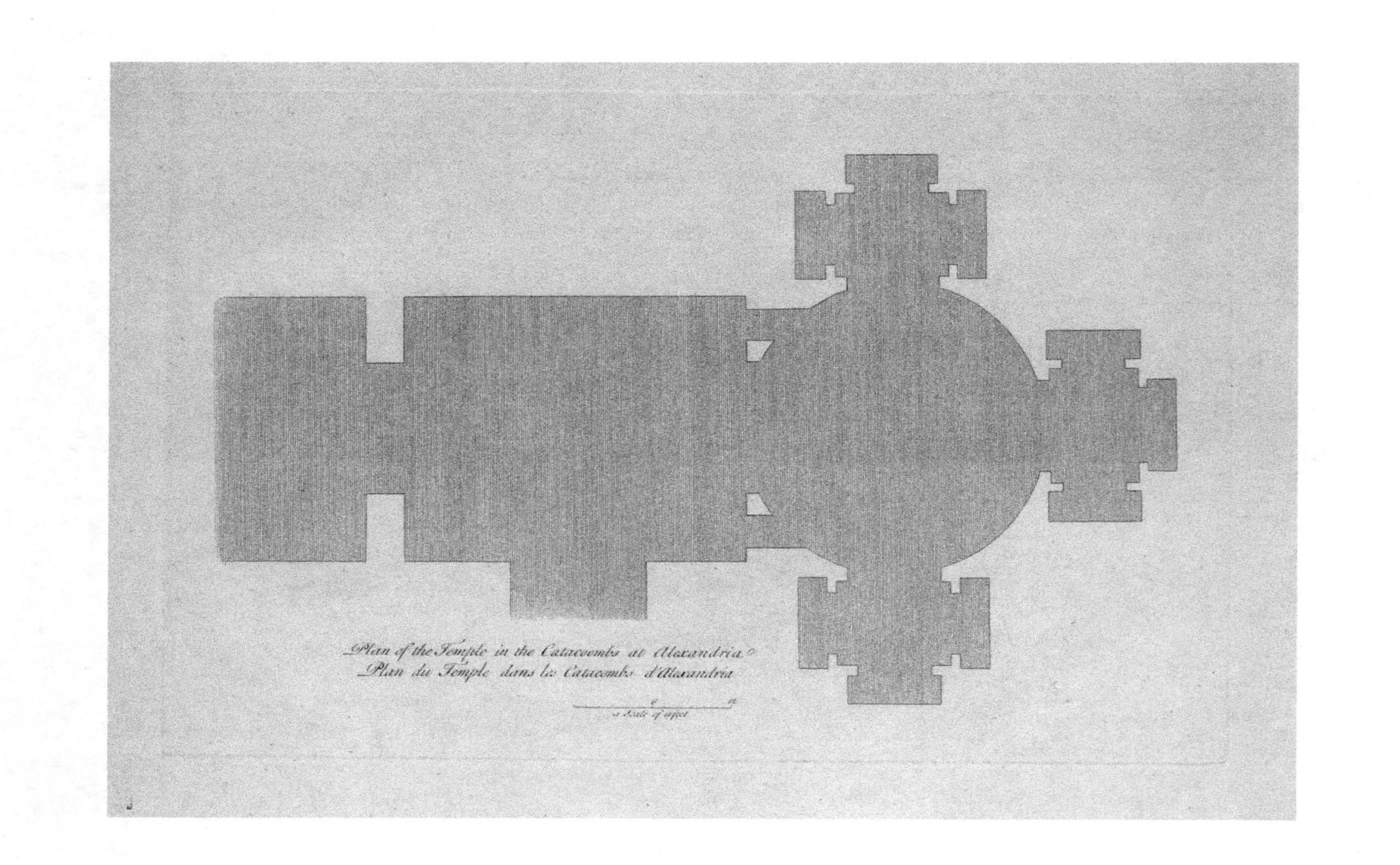

Plan of the Temple in the Catacombs at Alexandria
Plan du Temple dans les Catacombs d'Alexandria

The Large Granite Column at Alexandria (commonly called Pompey's Pillar) composed of only four pieces, the Capital one, the whole Shaft another, the Base and part of the Plinth the third, and the remainder of the Plinth with the Pedestal the fourth.
for part of an Inscription just discover'd by Dr. Pococke, Vid: Pococke on Egypt Vol. 1st P. 8th.

La Colonne de Granite a Alexandrie, (nommée La Colonne de Pompée) Elle ne consiste que de quatre Pierres, dont le Chapiteau en est une, le Fust de la Colonne en est une autre, le Base et partie de la Plinthe au dessous, en est une autre, et le Reste de cette Plinthe avec le Piedestal font la quatrieme.